ROAD & TRACK

ON

MG SPORTS CARS

1962-1980

Reprinted From
Road & Track Magazine

ISBN 0 946489 81 5

Published By
Brooklands Books with permission of Road & Track

Distributed By

Road & Track
1499 Monrovia,
Newport Beach,
California 92663, U.S.A.

Brooklands Book Distribution Ltd.,
PO Box 146,
Cobham, Surrey KT11 1LG,
England

Printed in Hong Kong

We are frequently asked for copies of out of print Road Tests and other articles that have appeared in Road & Track. To satisfy this need we are producing a series of books that will include, as nearly as possible, all the important information on one make or subject for a given period.

It is our hope that these collections of articles will give an overview that will be of value to historians, restorers and potential buyers, as well as to present owners of these automobiles.

We are indebted to R.M. Clarke for our cover photograph of an MGB taken in Australia.

MG-B 1800

Civilization has come to Abingdon-on-Thames

THE LOUDEST WAILS in the land were heard back in 1955 when MG made two shattering announcements: 1) the semi-classic TF was being dropped in favor of aerodynamics, and 2) the sturdy XPEG engine was to be replaced by an Austin "B" type. The body change was inevitable and was expected, we think, by most MG owners.

But for 7 years the streamlined MG-A consistently broke all previous sales records and proved in competition that it was faster and at least as durable as any of the good old T-series.

The new "B" isn't quite as much of a change as the "A" and no wails have been heard around our office. Our styling experts (who really are) never had much good to say about the lines of the A. It was "corny, out-of-date in 1955, had poor surface development, etc." But there's no complaint over the fresh new look for the B. The worst we heard was that it's good but not very advanced or exciting. Perhaps this is true if you're comparing it to, say, an Avanti, but everyone on our staff was enthusiastic over the appearance of the B, consultants notwithstanding.

Our enthusiasm did not wane during 700 miles of driving. In fact, it grew stronger, and frankly this is the first British car in several years which created no arguments among the staff—even the Italian and German sports car owners forgot their private battle and admitted they liked to drive this new English job.

The ride seems to be unchanged and this is both a fault and a virtue. There's no doubt about it, the ride begins to feel firm after an hour or two, yet it is this taut feel which gives the car its typical good handling in sharp corners or in fast bends. The steering characteristic is very close to being neutral at all times and when we weighed the car with full tank and driver aboard we found out why. The front wheels carried 1130 lb, the rears carried 1130 lb. (Actual curb weight was 2080 lb with wire wheels and radio.) The steering ratio is just a fraction slower and, we think, the better for it. The feel was a little less trigger-happy and more relaxing at high speed (2.9 turns lock to lock; formerly 2.6).

The larger 1796-cc engine feels happier too. Though the axle ratio has been dropped from 4.100 to 3.909 the new 14-in.

5

wheels negate any effect and 3340 engine revs are churning at 60 mph. Still, the engine seems smoother and quieter than the 1622-cc unit we tested two years ago. Part of this improvement may be attributable to the slightly lower compression ratio but we believe more efficient silencing is a more important factor. The 4 cylinders run like clockwork up to about 4000 rpm—above that there is a harsher note—but no vibration all the way up to the red line at 5500 rpm.

Performance recorded during the acceleration tests needs no apology—the B goes fractionally better than the A Mark II but there are only 4 more horses and the larger engine is obviously detuned slightly because an 11% increase in piston displacement gives only 4.5% more power. This may also be a factor in the smoother running mentioned earlier.

Incidentally, during the test runs it was very hot and the engine was barely broken in. We used 5500–5600 rpm as an absolute limit though 6000 can be used occasionally and such a speed would probably knock a tenth or two off the times we recorded. At any rate, and despite the above, the engine temperature never went above 170° F and at steady cruising

the needle held 160–165° F, during both day and night.

Fuel consumption worked out at 24 mpg around town and averaged 27 mpg on one 300-mile trip—a trip that was about evenly divided between slow traffic and fast freeway driving.

A big improvement was noted in the transmission. The A-types had good but rather stubborn synchronizers. The B, even with very few miles on it, shifted perfectly and noticeably easier. The shift lever seems to be a little higher than before—or actually the seats are just a little lower. A really worthwhile cockpit change comes via the new unit-construction which dispenses with the frame. There is, at last, ample pedal spacing for the average American shoe (foot?). The clutch and brake pedal pads are a little small (1.75 in. wide) but are spaced on 4.7-in. centers. Biggest improvement is accelerator pedal room—the space between the tunnel and the edge of the brake pad measures 6.1 in. The pedals didn't satisfy some of our drivers; either the brake was too high or the accelerator too low; in any case it was impossible to heel-and-toe, which is unfortunate but could easily be corrected.

The disc/drum brakes were absolutely without fault and despit

R&T ROAD TEST

MG-B 1800

SCALE: 10" DIVISIONS

DIMENSIONS

Wheelbase, in	91.0
Tread, f and r	49.2
Over-all length, in	153.2
width	59.9
height	49.4
equivalent vol, cu ft	262
Frontal area, sq ft	16.4
Ground clearance, in	4.2
Steering ratio, o/a	n.a.
turns, lock to lock	2.9
turning circle, ft	32.0
Hip room, front	2 x 18.5
Hip room, rear	n.a.
Pedal to seat back	45.0
Floor to ground	9.1

CALCULATED DATA

Lb/hp (test wt)	25.6
Cu ft/ton mile	88.4
Mph/1000 rpm (4th)	17.9
Engine revs/mile	3340
Piston travel, ft/mile	1950
Rpm @ 2500 ft/min	4290
equivalent mph	77.0
R&T wear index	65.1

SPECIFICATIONS

List price	n.a.
Curb weight, lb	2080
Test weight	2400
distribution, %	50/50
Tire size	5.60-14
Brake swept area	350
Engine type	4-cyl, ohv
Bore & stroke	3.16 x 3.50
Displacement, cc	1796
cu in	109.6
Compression ratio	8.75
Bhp @ rpm	94 @ 5500
equivalent mph	99
Torque, lb-ft	107 @ 3500
equivalent mph	62.8

GEAR RATIOS

4th	(1.00)	3.91
3rd	(1.37)	5.37
2nd	(2.21)	8.66
1st	(3.64)	14.2

SPEEDOMETER ERROR

30 mph	actual, 29.8
60 mph	58.5

PERFORMANCE

Top speed (5900), mph	106
best timed run	n.a.
3rd (5600)	73
2nd (5650)	46
1st (5650)	28

FUEL CONSUMPTION

Normal range, mpg	24/29

ACCELERATION

0-30 mph, sec	3.9
0-40	5.9
0-50	9.0
0-60	12.5
0-70	17.7
0-80	25.0
0-90	34.5
Standing ¼ mile	18.5
speed at end	72

TAPLEY DATA

4th, lb/ton @ mph	210 @ 58
3rd	300 @ 48
2nd	450 @ 35
Total drag at 60 mph, lb	105

ENGINE SPEED IN GEARS

ACCELERATION & COASTING

Removable hardtop will be available soon. Dished floor would allow inversion of spare, increasing luggage space.

MG-B 1800

there being no booster the pedal pressure was very moderate. Furthermore, there were no signs of some of the faults we often encounter. The action was progressive and produced no lurching as the car came to a dead stop. Of course, there is no fade either and we were surprised at finding none of the disc brake squeal which is all too common. The handbrake was adequate though it has a very short travel.

The instruments look to be the same but there is a cowl over the speedo and tach. The seats are now much better contoured and we got a surprise when we moved them all the way back. An average-height driver couldn't reach the pedals at all from this position and the critical pedal to seat back dimension is the longest we've ever encountered: 45.0 in. to the last hole setting and an inch more available past the last latching point. Two 6 ft 4 types did not need the seat all the way back and they still had headroom. Behind the seats is a carpeted shelf upon which an average size adult or two children can perch without too much discomfort.

The new doors measure 36.5 in. wide, making it easier to

get into the car. The top isn't quite so convenient. It folds neatly and flush into the well but the number of hinge points and top bows is confusing and it's easy to get pinched in the process. The wind-up windows (for the first time) are a great boon. They fit well, crank up or down with just over 2 turns and had a tendency to rattle on rough roads.

Our test car had a heater but there was no opportunity to use it. We were impressed with its neat and accessible installation with the business part on a shelf just behind the engine. The entire engine compartment is also well planned and the only tight spot we could find was the steering shaft coupling (at the pinion) where dual exhaust pipes, engine mounts, etc., preclude even finding it—let alone working on it.

Moving back to the rear, the higher rear fenders should give more trunk space, but volume is still skimpy and marred by the spare wheel and tire being smack in the middle. Our thinking is that the gas tank really should go somewhere else (maybe in the fender skirts like a Jaguar Mk X) and at least the wire wheel could be flopped over for more room.

Our test car was a very early production model and yet, when we went over every square inch, the quality of workmanship and lack of flaws were remarkable. This is the best engineered, the best put-together MG we've ever seen. ◎

Improvements in the Breed... MG for 1963

BMC makes spectacular advances on two fronts

IN NOVEMBER of 1955 the MG division of British Motors Corp. announced a completely new car, the MG-A. Only the TD-TF front suspension, slightly modified, carried over: the "A" had a new frame, new bodies, new engine, new transmission and new rear axle.

Now we have the MG-B-1800, replacing the MG-A-1600 Mark II. This time there is an all-new body, and with unit construction. Therefore, of course, no frame. Mechanical changes are minor; principally, the engine size has been increased from 1622 to 1798 cc, a gain of 11%.

It is interesting to note that the new MG-B is 3 in. shorter in wheelbase and overall length, and 2-in. wider than before. As a matter of fact, the box volume in cubic feet works out as exactly the same, at 262, while the frontal area is increased fractionally—by 2%.

The tabulation shows that the change to unit construction hasn't saved any weight (weights are with full tank, disc

wheels, no accessories). However, we must remember that the "B" has wind-up windows, a much roomier cockpit, etc. Also, the overall torsional rigidity is improved by unit construction and, though open-car cowl-shake was never a serious problem with the A, the net result is an even sturdier chassis, obtained with no increase in weight.

Unlike the Austin-Healey Sprite, which pioneered unit construction in BMC sports cars, the MG approach is entirely different. To explain a bit, the Sprite and its newer twin, the MG Midget, employ fairly simple flat section members wherever possible in order to simplify the tooling problem of (relatively) low volume production. But there is no sign of this type of economy in the MG-B. There are double rocker-panel sections under the doors. This member is straight and has overall dimensions of about 5.5-in. square—a generous size for what might be described as a frame rail. But in front there is a very elaborate sub-frame assembly designed to feed

EVOLUTION OF THE MG

M, 1929–32

J2, 1932–34

PB, 1935–36

front suspension loads into the ample cowl structure.

A somewhat similar sub-frame provides the rear kick-up, and supports the luggage compartment and the rear bumper brackets. The tunnel structure which surrounds the transmission and propelled shaft acts as a central backbone and contributes to the overall strength and rigidity. Outside sheet metal, particularly the front and rear fenders, also is stressed and therefore contributes its share towards overall stiffness. This entire structure is all-steel, but the engine hood is aluminum because it is rather large and lifts up (from the front) without the aid of helper springs. The luggage lid, being small, is steel, as are the doors.

While the tooling cost for such a structure is obviously high, the ultra-modern styling would seem to assure the continuance of this model for several years—perhaps even for the 7-year duration of the MG-A. In any case, the price is expected to be little more than before.

Mechanically, there are few changes. But the designers have somehow found room for yet another increase in cylinder bore. Remember, the BMC "B-type" engine started out in 1948 with the Austin A-40 and a cylinder bore of only 2.58 in. (1200 cc). Late in 1955 the bore was increased to 2.875 in., giving 1489 cc. With this change the block was completely redesigned and all crankshaft bearings were enlarged in diameter. When the twin-cam engine came along, the bore was increased to 2.96875 in. (1588 cc) and, more important, the crankshaft was redesigned to give thicker crank checks—necessary with the horsepower jump to 108 at 6700 rpm.

Then, in 1961 the ohv B-block got another bore-increase, this time to an even 3 in. and 1622 cc. This engine produced 90 bhp and it got the sturdier crankshaft from the twin-cam powerplant. (The stroke of this engine was 3.50 in. from the very beginning, and still is in all applications of the unit.)

For 1963 the designers have miraculously achieved what must surely be the last and largest possible bore change—an increase of $\frac{5}{32}$ of an inch, or 3.15625 in. to be exact (MG rounds it off to 3.16 in.). This gives 109.5488 cu. in. or 1795.87 cc by our calculations, though the catalog says 1798 cc and the above assumption may not be correct down to the last decimal place.

Whenever you increase the cylinder bore of a basic block from 2.48 to 3.16 in., something has to give. Obviously the cylinder bores in the casting are siamesed in pairs. This simply means that there is no water space between cylinders 1 and 2, and between 3 and 4 (i.e., the space where there are no main bearings, below). But some 3 million Model A Fords were built in this fashion and they certainly enjoyed an excellent reputation. And that was before the advent of cam-ground pistons. This type of piston makes the siamesed cylinder bores even more practical though it may be significant that the compression ratio has been reduced slightly on the B.

The transmission and rear axle are unchanged except that the synchro mechanism has been improved and the axle ratio has been dropped from 4.100 to 3.909:1. However, engine revolutions per mile are essentially the same as with the 1622-cc engine because 14-in. wheels replace the former 15-in. size.

There is an interesting change in the front suspension though essentially this assembly stems from the MG-Y sedan and the TD roadster of many years back. This i.f.s. scheme features a long steering-pivot with threaded bronze fittings on each end to allow steering movement. Suspension movement is obtained via cross-holes at 90° to the threaded axis—a sort of universal joint which has a great deal of merit and is still used on our Rambler. The upper bronze fitting on the MG-B is unchanged but the lower assembly has been redesigned to give better ground clearance.

TA/TB/TC, 1936–49

TD, 1950–53

TF, 1953–55

MG-A, 1955–62

LeMans MG-B

BY DAVID PALMER

With le mans over for another year and the Ferrari domination still unbroken, it was quite a pleasant surprise to see a famous name reappear in the class results.

The MG octagon, famous for its achievements on this arduous circuit before the war, was back, and the latest model, the MG-B, took 12th place overall, winning its class. This was the same car entered by the British Motor Corporation at Sebring, loaned to British driver Alan Hutcheson to enter the 24 hour marathon. His choice of partner was the well-known rally driver, Paddy Hopkirk.

The bucket seat and the starkness of the interior of this racing version at once made it obvious that this was a car far removed from the standard MG-B.

The dash panel had been cut in half and a large electric tachometer installed in place of the normal cable-driven instrument. A small-diameter wood-rimmed steering wheel had the MG motif in its center, and all carpeting and unnecessary trim to the rear of the seats were removed. The factory-built hardtop was reinforced over the driver's head by a solid-looking roll bar and the throttle pedal was of increased size to aid the toe-and-heel action of the driver.

The body itself was virtually all aluminum, the nose cone being of similar design to that used on the Dick Jacobs racing Midgets. In addition to the normal hood catch, two strong leather straps secured the hood in its closed position. At the rear, the luggage boot was taken up with a 20 (Imperial) gallon fuel tank, the quick-release filler of which protruded through the deck lid. Apart from the windscreen and quarter-lights, the other windows were all of plexiglass and, with all these modifications for lightness, the weight of the car was down to about 1900 lb from the normal 2070.

MG-B Number 7 DBL had commenced life on the normal production line and then had been turned over to the Competition Department. The engine was removed and completely dismantled, with the cylinder head receiving the usual at-tention: polished combustion chambers, gas-flowed ports, compression raised to 10.7:1, and careful matching of the inlet and exhaust manifolds to the ports. The bottom end (crankshaft, flywheel, clutch and con rods) was fully balanced and a high-lift camshaft, giving a greater valve overlap period, was installed. The cylinder bores remained standard, as did the main and big-end bearing shells. When all the parts had been attended to satisfactorily the unit was reassembled with loving care and the Weber 45-DCO carburetor on its special manifold installed. The exhaust manifold was a 3-branch free-flow special job feeding through to a large-bore pipe having a small Servais silencer fitted toward the rear, the exhaust outlet being just ahead of the near-side rear wheel.

The standard Borg & Beck 8-in. diaphragm-spring clutch was used, although the driven plate naturally had special hard-wear linings. During practice at Le Mans the gearbox front oil seal leaked and soaked the clutch; a new seal and clutch assembly had to be installed in double-quick time, and then no further trouble was encountered from this quarter during the whole 24 hours.

Brakes and suspension were as used on production models, except that the rear shock absorbers were of the adjustable type so that the stability could be altered to suit the driver's requirements. Ferodo DS-11 front disc pads were used and VG 95/1 linings at the rear. The optional extra wire wheels (available as such on the production models) were shod with Dunlop 5.00-14 D-9 racing tires, these giving excellent service during the race. One complete set was used, the two rears being changed at 12 hours, the two fronts at 18 hours. At this latter pit stop the front disc pads were also renewed.

Oil consumption during the race was in the region of two pints every 250 miles and gas was being consumed at one (Imperial) gallon every 13 miles. The drivers were using a rev limit of 6000 at all times, this being equivalent to about 131 mph down the straight in top gear with the low axle ratio. They were lapping the car regularly in 4 min 50 sec, or 104 mph (except for one excursion into a sand bank), during the whole race.

This is just one of the successes notched up by BMC's Competition Department, but it is one which probably counts for more prestige than any other race or rally in the motor sport calendar.

MGB GT COUPE

*And still the standard by which
other sports cars are measured*

IN THIS LATTER-DAY era of sports cars, getting into an MG again never fails to remind us of the Good Old Days when the MG, the TC in those days, was the first and only sports car available in America. Looking back over our road tests of the ensuing four basic models descended from the TC, and the several variations thereof, we find that MGs have been consistently fun to drive, tolerably well put together and outstanding value for the money. The latest MG is the MGB GT coupe, a closed version of the familiar MGB that has been with us since late 1962. Does the new coupe carry on the tradition?

We had expected an MGB coupe right from the introduction of the car, so it seems late getting here. Perhaps BMC hadn't really been serious about producing one until the current, international Fastback Fad got under way. If so, their answer to the fad is not exactly a "me too," and we're glad of that. Instead, MG body designers and stylists have come up with a fairly original-looking shape that goes well with the basic B roadster, makes good use of what room there was to work with and provides station-wagon-like utility for the occasional kitchen door that must be carried home from the lumber yard, what with 10 sq ft of cargo area with the jump seat folded down and a tailgate that can be left open. The luggage space figure in the data panel (4.5 cu ft) can be improved upon if one doesn't mind giving up his rear vision.

As for the jump seat itself, it could be best compared to that in the Austin-Healey: useless for adults, but as long as one doesn't plan to carry adults back there (well, one *could* sort of curl up) it should do nicely for a couple of small children and the tots will enjoy the cargo floor area as well. The driver of the MGB GT has benefited too, as the windshield and side windows are quite a bit deeper than on the roadster and combine with a generous rear window at a usable angle to provide very good driver vision. One staff member actually reaped a traffic ticket as a result of the low windshield header of the roadster when it prevented him from seeing a high-mounted "No Left Turn" sign at the exit of a drive-in restaurant.

Driving the coupe is not much different from driving the roadster but the car shows the results of three years of pro-

MGB GT COUPE
AT A GLANCE...

Price as tested	$3230
Engine	4 cyl inline, 1798 cc, 98 bhp
Curb weight, lb	2308
Top speed, mph	105
Acceleration, 0–60 mph, sec	13.6
50–70 mph (3rd gear)	8.1
Average fuel consumption, mpg	23

MGB GT COUPE

duction and detail development. Two of the major complaints against the earlier Bs have been eliminated. Engine noise and vibration, especially the latter, have been reduced noticeably, if not dramatically, by the 5-bearing crankshaft, rendering the 1800-cc engine one of the better in-line fours in these departments. And the rear axle, always chronically noisy with the MGB, has been quieted nicely by the substitution of a steel nosepiece for the previous aluminum one. These changes combine with the added sound insulation of the coupe and its decent wind noise level to make one of the quieter-running small sports-touring cars we've driven; however, the smoother engine hasn't completely eliminated the old, disturbing resonance points and one still avoids the 4000-rpm neighborhood for steady cruising—most unfortunate, as that's 70 mph. In our opinion, the optional overdrive ($185) would be almost essential, and certainly desirable.

Another small, but annoying, problem has remained with the MGB right up through our test car: a curious jingling noise that originates in the starter Bendix drive and reminds one when 2500 rpm is reached. It's even more curious that BMC has allowed this problem to remain so long.

Suspension still consists of conventional A-arms at front and a live axle on leaf springs at the rear; the ride is still firmer than it might be with an independent rear, but as the car's performance level isn't very wild it has less need than some cars for the more sophisticated suspension linkage. The body structure certainly does its part admirably—the B roadster has been one of the most rigid open cars around and the coupe improves to the point of being Gibraltar-like. Very pleasant in this department.

Steering is a strong point, as it always has been with MGs Still positive, light enough (strong caster keeps it from being as light as it could be) and without a trace of free play, it's a joy. And the B is about as insensitive to side winds as they come—like a Toronado, in fact. The anti-roll bar has been made standard on the coupe, and hence roll is less noticeable than on the last B we tested; no bother here at all, though it's greater than it was with the TC.

If we ever thought MGs had outstanding brakes, perhaps it was because our tests weren't severe enough. The panic stop from 80 mph netted as a respectable deceleration rate of 26 fps^2 with good control in spite of some late rear lockup; but the newly instituted fade test, not a drastic one, brought out the worst in the B's disc-drum setup. They faded to the point of a 50% (roughly) increase in pedal pressure by the sixth stop from 60 mph at 0.5 G, and on the last few feet of that sixth stop they were just about gone. They were also quite slow to recover from this faded condition.

Our most comfort-conscious staff member found the seats acceptable but wished for extra thigh support for the long drive. Seat belts are necessary if one is to remain in place while cornering. We appreciated the rather slick leather of the seats, for it makes for easy in- and out-getting and for not pulling out your shirt tail. The seats are adjustable so far back that the longest legs will fit, if not the tallest head. Plastic door pulls are provided, but we prefer armrests in their place. A staff member found a pair of rear door armrests from a late Mercury that mounted easily and looked appropriate to the MG's restrained interior, and installed them on his MGB.

Ventilation has been provided with a cowl fresh-air intake and a large two-position inside flap that has a knuckle-smashing handle. The heater, standard now on the coupe, is probably just adequate for the closed car, but it's sub-marginal for the roadster in northern winters. Only one blower speed is provided. MGs generally come equipped with 160° thermostats and it's very wise to use about a 185° 'stat for the winter, if yours are severe.

As an everyday, livable and lovable car, the B is hard to equal. It starts readily if not instantly down to 0°F (heaven help you below that), runs smoothly at idle and low speeds, and has a general tractability that is legend with MGs. The major controls are handy, especially the gearshift and handbrake and even those who grow weary of manually shifted cars that really need to be shifted won't mind this job much. Heel-and-toeing is possible but not especially easy. One staff member thinks BMC would be wise to offer an automatic transmission one of these days.

Minor controls, however, are less satisfactory. Three toggle switches side by side operate the lights, wipers and heater blower, and they're not labeled; the only interior light is the little map light on the right end of the dash. It's difficult to adjust the heater at night, as there's no light for the controls and they are some distance from the driver. Better leave this to a passenger. We did appreciate the headlight flasher, operated by pulling the directional stalk rearward—this is an item we'd like to see universalized.

Instrumentation is firmly in the MG tradition with adequate-sized speedometer and tachometer (now electric) side by side, beautifully marked and well lighted at night. One point of entertainment has been lost, though: the fuel gauge has finally got some damping, so that it gives a steady reading. MG fuel gauges of the past have been somewhat more useful as accelerometers than fuel gauges, and we were convinced at first that something was wrong with the one on our test car. The water temperature gauge on MGs has been a trouble spot in the past too, and will continue to be as long as BMC uses the type of instrument 'seen on the MGB. It has a long coil of soft metal tubing leading from a bulb in the water passage, the coil containing some sort of fluid that expands with heat to give readings on the gauge. This tube gets knocked around and bent by mechanics working on the engine, in addition to just vibrating with the engine, and invariably fatigues and breaks. Most cars use sending units these days—that is, the reading is transmitted to the gauge by an electrical wire. The oil pressure gauge is also a direct instrument in the B and must be replaced with the temperature gauge as they share the same housing.

As before, we must complain about the gearbox itself—but not the excellent linkage. This was an outstanding unit in the late 1940s, but now we have come to take a synchronized first gear for granted, and still the B doesn't have it—even though its sister the Austin 1800 and most of its direct competitors (Alpine, TR-4A, Fiat 1500, Datsun 1600) do. There are plenty of situations where 1st is needed while still rolling, and the compromise of 2nd gear ratio necessitated by the lack of synchromesh greatly reduces the potential usefulness of that gear (See Speeds in Gears). The old bugaboo of getting into 1st at rest remains too; we've seen many an MG almost get run over at a green light when those gear teeth didn't stop in the right place.

As curb weight is up by 228 lb over our last B and power up only 4 bhp, we would have expected some loss of performance with the coupe body. The loss is mild: the coupe takes 1.1 sec longer to reach 60 mph (13.6 vs. 12.5 sec) and the same amount longer to cover the quarter mile (19.6 vs. 18.5).

Everything about this latest MG says, Yes, the MG tradition had been upheld. It is fun, it has a pleasant personality, it's put together and finished properly, and we think it remains one of the best values in its field. It could be more modern but it could be less, too—long live the king.

BMC B-series engine has reached a high state of refinement and reliability after 11 years' development. Engine access is good.

Engine oil cooler, standard on all MGs sold in the U.S., contributes significantly to engine life at high cruising speeds.

ROAD TEST
MG B GT COUPE

SCALE: 10" DIVISIONS

PRICE

Basic list................$3160
As tested................$3230

ENGINE

No cyl & type.......4 inline, ohv
Bore x stroke, mm....80.3 x 89.0
  In...............3.16 x 3.50
Displacement, cc/cu in.1798/110.5
Compression ratio..........8.8:1
Bhp @ rpm...........98 @ 5400
  Equivalent mph............107
Torque @ rpm, lb-ft..107 @ 3500
  Equivalent mph.............61
Carburetors............2-SU HS4
  No. barrels, dia.......1 x 1.50
Type fuel required......premium

DRIVE TRAIN

Clutch type.......diaphragm sdp
  Diameter, in..............8.0
Gear ratios: 4th (1.00).....3.91:1
  3rd (1.37)............5.35:1
  2nd (2.26)............8.82:1
  1st (3.65)...........14.2:1
Synchromesh........on top 3
Differential type..........hypoid
  Ratio..................3.91:1
  Optional ratios...........none

CHASSIS & SUSPENSION

Frame type.......unit with body
Brake type.........disc/drum
  Swept area, sq in.........239
Tire size.................5.60-14
  Make.................Dunlop
Steering type......rack & pinion
  Turns, lock-to-lock.........2.9
  Turning circle, ft..........32
Front suspension: independent with
  unequal A-arms, coils, lever
  shocks, anti-roll bar.
Rear suspension: live axle, leaf
  springs, lever shocks.

ACCOMMODATION

Normal capacity, persons.......2
  Occasional capacity..........3
Seat width, front, in.....2 x 20.0
  Rear................33.5
Head room, front/rear..32.6/26.0
Seat back adjustment, deg......5
Entrance height, in........45.5
Step-over height..........13.7
Door width.................34.0
Driver comfort rating:
  Driver 69 in. tall..........95
  Driver 72 in. tall..........90
  Driver 75 in. tall..........85
  (85–100, good; 70–85, fair; under
  70, poor)

GENERAL

Curb weight, lb............2308
Test weight................2683
Weight distribution (with
  driver), front/rear, %..48.5/51.5
Wheelbase, in.............91.0
Track, front/rear.....49.2/49.2
Overall length...........153.2
  Width.................59.9
  Height................49.8
Frontal area, sq ft.........16.4
Ground clearance, in........4.5
Overhang, front/rear....26.8/35.4
Departure angle, deg........20.3
Usable trunk space, cu ft.....4.5
Fuel tank capacity, gal......12.0

INSTRUMENTATION

Instruments: 120-mph speedometer, 7000-rpm tachometer, fuel, oil pressure, water temperature. Warning lights: ignition-generator, directionals, high beam.

MISCELLANEOUS

Body styles available: coupe as tested, roadster.
Warranty period: 12 mo./12,000 mi

CALCULATED DATA

Lb/hp (test wt).............27.3
Mph/1000 rpm (4th gear)....17.5
Engine revs/mi (60 mph)....3430
Piston travel, ft/mi.......2000
Rpm @ 2500 ft/min.......4290
  Equivalent mph.........78
Cu ft/ton mi..............82.2
R&T wear index............68.8

EXTRA COST OPTIONS

Overdrive, whitewalls, chrome wheels, radial or Road Speed tires, radio*, cigarette lighter.
*Fitted on test car.

MAINTENANCE

Crankcase capacity, qt........5.0
  Change interval, mi........6000
Oil filter type...........full-flow
  Change interval, mi........6000
Chassis lube interval, mi.....3000

ROAD TEST RESULTS

ACCELERATION

Time to speed, sec:
  0–30 mph................4.0
  0–40 mph................6.3
  0–50 mph................9.9
  0–60 mph...............13.6
  0–70 mph...............18.3
  0–80 mph...............25.1
  0–90 mph...............37.2
  50–70 mph (3rd gear).......8.1
Time to distance, sec:
  0–100 ft.................3.6
  0–500 ft................10.3
  ¼-mile.................19.6
Speed at end, mph...........72
Passing exposure time, sec:
  Car ahead going 50 mph.....7.9

FUEL CONSUMPTION

Normal driving, mpg......20–26
Cruising range, mi......240–312

SPEEDS IN GEARS

4th gear (5300), mph........105
3rd (6000)................81
2nd (6000)................47
1st (6000)................29

BRAKES

Panic stop from 80 mph:
  Deceleration, % G..........80
  Control.................good
Parking: hold 30% grade.......no
Overall brake rating........good

SPEEDOMETER ERROR

30 mph indicated......actual 28.5
40 mph.................37.2
60 mph.................56.4
80 mph.................77.1
90 mph.................87.7
Odometer correction factor...0.986

ACCELERATION & COASTING

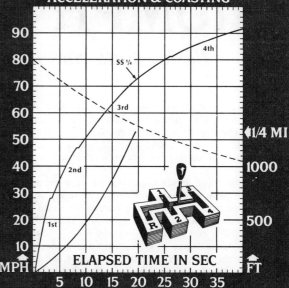

ELAPSED TIME IN SEC

GORDON CHITTENDEN PHOTOS

MG MIDGET III
TRIUMPH SPITFIRE Mk 3

Basic sports cars that continue a proud tradition

THE MG MIDGET and the Triumph Spitfire are basic sports cars, the type most likely to mark the driver's first ownership of the breed. As such, they meet certain classic requirements and have a set of special traits that set them apart from their more sophisticated (and expensive) brethren.

One of the more subtle requirements of a proper sports car is that it have a proud name earned in road racing competition. Most manufacturers have learned the value of building a proper reputation of this type and those that don't almost inevitably end up wondering why their products never achieve the same degree of acceptance and respect. There's no question about the pedigree of the Midget and Spitfire in this field and the driver never has to suffer that vague unease that results from not being sure whether his car is or isn't a true sports car.

They also meet the most important all-round requirement for a sports car—responsiveness. If a sports car is anything, it is responsive—with steering that is quick and accurate, clutch action that is crisp, a gearbox with positive feel, brakes that can be used hard and often. All-out brute performance is far less important than the feeling that you are in complete control and directing the machine rather than holding on and trying not to do anything foolish.

There are still more characteristics. In a basic sports car there should be an awareness of things mechanical going on under hand, foot and seat. A bit of row under the hood when the accelerator is depressed, for instance; a businesslike exhaust note; even a little gear whine isn't unacceptable. There should also be instruments. Before their first sports car, most drivers have never had the pleasure of knowing a full set of instruments at work, felt the satisfaction of just kissing the

Open seam behind Midget wheel remains from days when whole front end lifted. Functional lid now starts at grille.

Raised front bumper on Spitfire will take getting used to. Entire front end swings open for ultimate in availability.

redline with the tach needle, or the doubt and dread that come with the rise of the water temperature and the accompanying descent of the oil pressure. Warning lights, in comparison, are as sterile as a dead battery.

Perhaps not strictly necessary but nonetheless acceptable in the basic sports car are certain minimum standards of habitability. They don't have to offer armchair comfort for the occupants or boxcar volumes of luggage space. And even something out of date, like a non-synchromesh first gear, can be tolerated simply because it makes the driver a part of a great tradition. It's appalling to think there are millions of drivers who've never known the satisfaction of a perfectly executed double-clutched downshift into a non-synchro first gear.

It isn't easy to think of two cars that fulfill all the requirements for the basic sports car any better than the MG Midget and the Triumph Spitfire.

THIS ROAD TEST is of the latest version of each of these models, the Midget Mk III and the Spitfire 3. (The Roman and Arabic designations are those of the manufacturers, by the way.) The major change for both cars is that each now has a 1.3-liter engine. As you are no doubt aware, the MG Midget differs from the Austin Healey Sprite only in name and trim. The original Sprite came along in 1958, a stark 1-liter sports car with frog-eye headlights on the hood. The current body style was introduced in 1961 along with a 1098-cc engine and disc brakes at the front. The Midget nameplate

made its debut at the same time so that the second-series Sprite was the original-series Midget and the Midget has been one number behind ever since. In 1964, the Midget II adopted roll-up windows and semi-elliptic rear springs to reduce the tricky roll oversteer inherent in the live-axle-on-quarter-elliptics design.

The new Midget III's 1275-cc engine is similar to that used in the Mini Cooper S but is detuned from 75 bhp to 65 which allows lower production costs (a normal forged crankshaft can be used, for example, instead of the more expensive nitrided steel crank of the S) and yet continue the Midget's reputation for reliability and long life. There is a net increase of 6 bhp over the Mk II's 59, though, so the performance is somewhat better.

The other major change in the Midget III is the new top. This is now a proper convertible top that goes up and down easily and accurately and is a great improvement over the roadster-style build-it-yourself top.

The Triumph Spitfire has a somewhat shorter history than its opposite number from BMC. It was introduced in 1962, underwent minor revisions and a power increase (from 63 to 67 bhp) in 1964 and is now offered in $3000 fastback form as the 95-bhp, 2-liter GT6 as well as the basic roadster with the 75-bhp, 1296-cc 4-cyl engine. It is somewhat more modern than the Midget in one respect as it has independent suspension at both ends while the Midget has a live rear axle.

The 1296-cc engine of the 3 is based on the 1300 Triumph instead of the 1147-cc unit used before. This engine has in-

MG MIDGET III
ROAD TEST RESULTS

PRICE
List price $2255
Price as Tested 2434

ENGINE & DRIVE TRAIN
Engine, no. cyl, type. inline 4, ohv
Bore x stroke, mm .... 70.6 x 81.3
Displacement, cc/cu in .. 1275/77.5
Compression ratio 8.8:1
Bhp @ rpm .... 65 @ 6000
   Equivalent mph 90
Torque @ rpm, lb-ft .. 72 @ 3000
   Equivalent mph 46
Transmission type ... 4-spd manual
Gear ratios, 4th (1.00) 4.22:1
   3rd (1.36) 5.73:1
   2nd (1.92) 8.09:1
   1st (3.20) 13.5:1
Synchromesh .... on top 3
Final drive ratio 4.22:1

GENERAL
Curb weight 1560
Weight distribution (with
  driver), front/rear, % 50/50
Wheelbase, in 80.0
Track, front/rear 46.3/44.8
Overall length 137.4
  Width 56.5
  Height 48.6
Frontal area, sq ft 15.3
Steering type .... rack & pinion
  Turns, lock-to-lock 2.3
Brake type, f/r ...... disc/drum
Swept area, sq in 190

ACCOMMODATION
Seating capacity, persons 2
Seat width .... 2 x 17.5
Head room 39.0
Seat back adjustment, degrees .. 0
Driver comfort rating (scale of 100):
  For driver 69 in. tall 70
  For driver 72 in. tall 60
  For driver 75 in. tall 45

PERFORMANCE
Top speed, high gear, mph 93
Acceleration, time to distance, sec:
  0–100 ft 4.0
  0–250 ft 6.7
  0–500 ft 10.5
  0–750 ft 13.8
  0–1000 ft 16.5
  0–1320 ft (¼ mi) 19.9
  Speed at end, mph 69
Time to speed, sec:
  0–30 mph 4.3
  0–40 mph 6.7
  0–50 mph 10.2
  0–60 mph 14.7
  0–80 mph 31.0

BRAKE TESTS
Panic stop from 80 mph:
  Deceleration rate, % g 81%
  Control .... good
Fade test: percent of increase in
  pedal effort required to maintain
  50%-g deceleration rate in six
  stops from 60 mph 60%
Overall brake rating .... good

SPEEDOMETER ERROR
30 mph indicated ..... actual 29.2
40 mph 38.8
60 mph 58.0

CALCULATED DATA
Lb/hp (test weight) 29.4
Cu ft/ton mi 87.2
Mph/1000 rpm (high gear) ... 15.4
Engine revs/mi 3900
Piston travel, ft/mi 2080
Rpm @ 2500 ft/min 4690
  Equivalent mph 71
R&T wear index 81
Brake swept area, sq in/ton ... 200

FUEL
Type fuel required ...... premium
Fuel tank size, gal 7.5
Normal consumption, mpg ... 23–25

ACCELERATION & COASTING

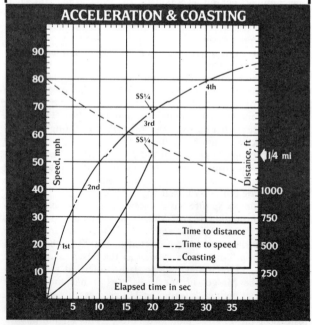

Speed, mph / Distance, ft / Elapsed time in sec

Legend: Time to distance / Time to speed / Coasting

Midget interior is small and snug; neat top is great improvement over earlier version; engine now has 1275 cc and 65 bhp.

dividual inlet ports instead of siamesed, the combustion chamber shape is improved and there is a cast iron manifold in place of the fabricated headers used on the Spitfire 2. Other mechanical changes include slightly larger brake calipers which should contribute to longer pad life and a new lift-over-and-clamp top that is superior to the old one in every way.

There are also changes in appearance that distinguish the 3 from earlier versions. The front bumpers have been raised to a more practical height, back-up lights are standard and the sharp-eyed will note that the exhaust pipe is slightly larger in diameter and now terminates at the right rear instead of the rear center. In the cockpit there is a smaller steering wheel, which gives the driver a bit more leg room, and there is a nice polished-wood setting for the instruments.

In size, the Spitfire is bigger than the Midget, longer in wheelbase and overall length (83.0 and 147.0 vs. 80.0 and 137.3) and heavier in curb weight (1680 vs. 1560 lb). In straight-line performance, the Spitfire is the quicker of the two, getting through the standing quarter in 19.3 to the Midget's 19.9 sec and having an edge in top speed of 100 to 93.

Both cars handle very well. The Midget is well balanced with a 50/50 weight distribution (compared to 54/46 front/rear for the Spitfire) and there is still a dependable bit of roll oversteer built into the rear suspension that makes it great fun to drive a little bit sideways. Predictably, the different rear suspension systems of the two cars makes for somewhat different handling. The overall effect is that the Spitfire rides and handles better over uneven road surfaces, as it should with independent rear suspension, has less initial oversteer with comparable steering effort but transfers to final oversteer more abruptly—again as you would expect with swing axles.

From the driver's point of view, the Midget seems to be better assembled, to have fewer rattles and to be more of a piece. The engine is smoother than the Spitfire's and the controls all seem to be happier with each other. For example, the Spitfire's throttle linkage has an "over-center" feel which requires a delicate touch to make small changes in throttle opening while the Midget's throttle action blends in unobtrusively well with the operation of the other controls. The Spitfire also has a slightly skewed shift pattern that takes

Spitfire has new wood-finished instrument panel; engine is uncommonly accessible; interior is roomier than Midget.

some getting used to while the Midget's shift lever is right where it should be and works just the way it should. The Midget is also quieter in everything except first gear and the engine seems willing to wing along at any reasonable speed without complaint. We did notice that after several miles of only moderately high cruising speed (70) in the Midget the oil pressure dropped from 60 to 45 psi, which suggests that the addition of an oil cooler might not be a bad idea if long periods of hard running are anticipated.

The Spitfire engine is rougher than the Midget's, noisier at low revs and with more vibration at speed. Triumph does offer overdrive on the Spitfire (which BMC does not offer on the Midget) and that would make touring more pleasant as no one really enjoys hearing a rough little 4 buzz out more than 4000 revs per mile at 70 mph.

The luggage and storage space is roughly the same for both cars. The spare tire and tools occupy a good share of the space in the trunks of both but the odd-shaped area left there and behind the seats is sufficient for even serious touring if you don't demand hard-sided luggage for your pliable belongings. Both cars are extremely agile for traffic and parking, the very tight (24-ft) turning circle of the Spitfire permitting close-quarter maneuvers almost unparalleled in modern motoring. The heater of the Midget is, frankly, abominable and there is no provision for bringing fresh air into the cockpit except through the heater, whose hot-water valve is reached only by looking under the hood. The Spitfire's heater, though also with a one-speed blower, is better and there are two underdash vents for fresh air.

There are many virtues in the Midget's small bulk—such as easy, accurate handling, confident maneuvering in tight quarters—but there are also several penalties to be paid. The driver compartment is so small as to be barely satisfactory for a 5-9 driver (see Driver Comfort Rating) and all but impossible for any driver much over 6 ft tall. The Spitfire is much more satisfactory in this respect.

It must be noted, though, that for comparably equipped cars, the Midget costs about $200 less. And for most drivers getting into their first sports car, $200 is a serious quantity of money. But whichever one the buyer chooses, he is assured of many miles of motoring pleasure in the great sports car tradition. They're good cars, both of them. You can't go wrong.

TRIUMPH SPITFIRE Mk 3
ROAD TEST RESULTS

PRICE
List price.................$2373
Price as tested...........$2672

ENGINE & DRIVE TRAIN
Engine, no. cyl, type..inline 4, ohv
Bore x stroke, mm.....73.7 x 76.0
Displacement, cc/cu in...1296/79.2
Compression ratio.........9.0:1
Bhp @ rpm...........75 @ 6000
  Equivalent mph............101
Torque @ rpm, lb-ft...75 @ 4000
  Equivalent mph.............64
Transmission type...4-spd manual
Gear ratios, 4th (1.00).....4.11:1
  3rd (1.39)............5.73:1
  2nd (2.16)............8.87:1
  1st (3.75)...........15.4:1
Synchromesh...........on top 3
Final drive ratio..........4.11:1

GENERAL
Curb weight, lb.............1680
Weight distribution (with
  driver), front/rear, %....54/46
Wheelbase, in..............83.0
Track, front/rear.......49.0/48.0
Overall length............147.0
  Width..................57.0
  Height.................47.5
Frontal area, sq ft.........15.1
Steering type......rack & pinion
  Turns, lock-to-lock.........3.8
Brake type, f/r.......disc/drum
  Swept area, sq in.........205

ACCOMMODATION
Seating capacity, persons.......2
Seat width.............2 x 19
Head room................39.0
Seat back adjustment, degrees...0
Driver comfort rating (scale of 100):
  For driver 69 in. tall........85
  For driver 72 in. tall........75
  For driver 75 in. tall........70

PERFORMANCE
Top speed, high gear, mph....100
Acceleration, time to distance, sec:
  0–100 ft.................3.8
  0–250 ft.................6.6
  0–500 ft................10.3
  0–750 ft................13.4
  0–1000 ft...............16.1
  0–1320 ft (¼ mi).......19.3
  Speed at end, mph.........70
Time to speed, sec:
  0–30 mph.................4.1
  0–40 mph.................6.1
  0–50 mph.................9.3
  0–60 mph................13.6
  0–80 mph................28.0

BRAKE TESTS
Panic stop from 80 mph:
  Deceleration rate, % g.......74
  Control..................good
Fade test: percent of increase in
  pedal effort required to maintain
  50%-g deceleration rate in six
  stops from 60 mph.......15%
Overall brake rating.....very good

SPEEDOMETER ERROR
30 mph indicated.....actual 27.3
40 mph........................37.5
60 mph........................57.5

CALCULATED DATA
Lb/hp (test weight).........27.0
Cu ft/ton mi................85.1
Mph/1000 rpm (high gear)....15.9
Engine revs/mi............3760
Piston travel, ft/mi........1880
Rpm @ 2500 ft/min........5000
  Equivalent mph............82
R&T wear index............70.8
Brake swept area, sq in/ton...202

FUEL
Type fuel required.......premium
Fuel tank size, gal...........9.9
Normal consumption, mpg...22-24

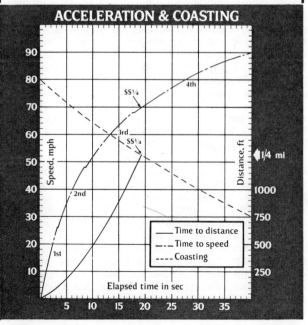

ACCELERATION & COASTING

After the New Wears Off:
1964 MGB ROADSTER 45,000 MILES LATER

BY JAMES J. EWING

WHETHER BOUGHT NEW or used, a low- or medium-priced British sports car has been the first purchase of many young enthusiasts. Most people base such a venture on the opinions of friends who know something about imports, the good word of the car salesman, and road tests published in the motoring press. However good a given road test may be, it is by its very nature limited in scope and really gives the prospective buyer no accurate way of projecting the cost of owning and operating such a vehicle over an extended period of time. Hopefully, this report will help fill this information gap for one of the more popular medium-priced sports cars, the MGB.

The most common question asked of the owner during his almost 3-year ownership is, "Why didn't you buy a Triumph?" The answer is one of personal taste. I like the looks and detail finish of the B much better than its competitors. Furthermore, I got a better deal on an equipped B than I could have gotten on a similarly equipped TR-4. Specifically, the car was purchased in July 1964 from Al Pernett's in Riverside, Calif., for a total price of $3044.66, which included state sales tax and license. Of the initial cost, $1700 was bank-financed over an 18-month period at a total cost of $189.46. The car was equipped with the following options: high compression engine (not standard in England), wire wheels, heater, tonneau cover, fold-away top, and front anti-roll bar. Overdrive would have been an excellent option and I have many times regretted that my car was not so equipped.

The car has been driven in an extremely wide variety of conditions including local commuting and high speed cross-country trips, in the snow of Chicago in winter and the heat of the Mojave Desert in summer. Up to about 30,000 miles the performance was excellent but the last 15,000 have seen some very discouraging failures. Gas mileage has always been superb and seems to get better with age. Radial-ply tires seem to have aided the cause of good gas mileage. Driving around

1964 MGB
Costs for 45,000 miles

Repairs & Replacements:

Broken steering wheel, 9500 mi..warranty	
4 recapped tires, 14,000 mi.	$72.26
Wheel alignment, 15,000 mi.	15.20
Front brake pads, 14,000 mi.	26.17
Front brake pads, 30,000 mi.	26.17
5 Pirelli Cinturatos & tubes, 31,000 mi.	177.30
2 6V Batteries, 34,000 mi.	47.30
Generator, 34,000 mi.	48.10
Rear axle seal, 37,000 mi.	12.25
Fuel pump, 38,000 mi.	43.23
U-joints, 40,000 mi.	44.25
Carburetor rebuilt & smog control service, 42,000 mi.	28.22
Exhaust system (see text)	65.00
Total	$605.45

Operating Costs:

Normal maintenance	$387.54
Gasoline (including oil between normal changes)	576.11
Total operating costs	$963.65
Cost per mi for 45,000 mi.	2.13¢

Overall Costs:

Delivered price	$3044.66
Financing charges	189.46
Operating costs	963.65
Repairs & replacements	605.45
License fees	135.00
Insurance	1243.00
Total expense	$6181.22
Less estimated value at end of period	$1200.00
Cost of driving 45,000 mi.	$4981.22
Overall cost per mile	11.07¢

CONTINUED ON PAGE 26

ALL-SYNCHRO MGB

Yes, it really is true,
the MGB has a fully synchronized gearbox

THE MGB has been with us for over five years now and probably has two more years to go. Truly British, its character is now vintage. It has finally been updated by the substitution of an all-synchromesh gearbox (over chief engineer Alec Issigonis' dead body?), and this, plus certain changes necessitated by air-pollution and safety legislation in the U.S., offers the opportunity to give it a fresh evaluation. An automatic transmission model will be along shortly too, and we will report on that when it is available.

The principal mechanical change is the new gearbox. It is marked visually by a straight gear lever instead of the bent one used before, and it sprouts from an opening farther back on the gearbox tunnel—perhaps a bit too far back, a criticism we almost never have to make. Shifting is much the same as before—a bit stiff and notchy, but wonderfully precise—and the new box not only eliminates the traditional 1st-

ALL-SYNCHRO MGB
AT A GLANCE

Price as tested	$2947
Engine	4 cyl, inline, ohv, 1798 cc, 92 bhp
Curb weight, lb	2220
Top speed, mph	104
Acceleration, 0–¼ mi, sec	18.7
Average fuel consumption, mpg	24.0

Summary: Strong performance, good durability . . . body, handling behind modern standards . . . synchro 1st gear, new indirect ratios big improvements . . . safety & emission changes not well worked out.

ALL-SYNCHRO MGB

gear whine but improves general drivability by virtue of its better ratios. Specifically, 1st gear is taller than before (at 13.45 overall vs. 14.2) and so is 2nd (8.47 vs. 8.82); 3rd and 4th are essentially the same, and a cruising ratio can still be added by ordering overdrive at $175. In short, it's now possible to drive the MGB like any other car, and it's a pleasant change.

The manufacturer has taken the air-injection route to controlling exhaust emissions. By virtue of the concomitant changes to carburetion and timing, peak torque is slightly *up* from 107 lb-ft @ 3500 rpm to 110 @ 3000, and peak power is *down* from 98 bhp @ 5400 to 92 @ 5400. The performance figures reflect the changes exactly: acceleration times are generally a little better than the early model we tested, but top speed is down from 106 mph to 104. Fuel economy is down somewhat, from an average of 26 mpg to 24. The pleasant exhaust note turns into a burble-burble on deceleration but we got only one definite backfire during our 500-mi test. One common side effect of air injection, a tendency to return to idle slowly, was present in our test car to the extreme—it took 9 sec for the engine to get back to idle from 3000 rpm after releasing the throttle! Otherwise, the 1798-cc pushrod four is pretty much the same: mechanically smooth, now that it has a 5-bearing crankshaft, but noisy; and furnishing abundant torque for the 2200-lb car. An alternator has replaced the generator and the old rattling starter drive has finally been cured.

Safety legislation has resulted in an entirely new dash panel for the MGB, plus a rearrangement of most controls. There is a molded, padded facia which attaches to the existing sheet metal, eliminating the glove box on the right and forming a new instrument cluster on the left. All instruments are now directly in front of the driver, and smaller—but still readable—speedo and tach have been substituted. The vague, poorly marked heater controls are also relocated, and all dash switches are non-protruding rocking tablets. Wipers are now 2-speed, as required by law, and are controlled by a right-hand stalk on the steering column; where legal, the left-hand directional lever also works a headlight flasher.

For the first time in history an MG roadster has sun visors, again thanks to the safety regs, and the recessed inside door handles are a good solution. Over-the-shoulder belts come from the rear deck panel and fasten, with the lap belts, by means of Kangol magnetic latches which we found convenient to use.

The seating position bespeaks the vintage character of the

car perhaps as much as the high noise level. One sits very low in the car, on relatively soft seats; the steering wheel is huge (perhaps because the steering is so stiff), the pedals are close together, and tall drivers rest their right shins against the center console. The hood and door sills are high, the windshield header low—but overall vision isn't bad, thanks to the 3-piece rear window. Sealing of windows to the top is indifferent, and it's not difficult to see daylight beyond the door edges. There are no door stops to keep them open on a slope, and the roadster top (either the standard put-away affair or the optional folding one) is pure British blacksmithery—for an example of how a good roadster top is designed in AD 1968, see the Fiat 124 Spider. The trunk is an example of what a storage compartment ought not to be, unlined and mostly occupied by the spare wheel and tire. The reader should bear in mind that many of these criticisms don't apply to the MGB/GT—but it costs a few hundred dollars more than the roadster.

Handling, too, is vintage, but if one accepts the fact, it is still possible to enjoy driving the MGB hard. Steering is very heavy, presumably from lots of caster; stiff springing and a live rear axle mean some hopping about on rough roads, but on smooth surfaces one can enjoy the ease of breaking the rear end loose and the relatively flat cornering. Our car had the now-standard anti-roll bar, which helps keep the body flat and still doesn't add too much understeer to the picture; this used to be an option.

Braking, like steering, is heavy in the MGB—the initial half-g stop in our fade test took all of 60 lb pedal effort. The disc/drum brakes bring the B to a smooth, controllable stop at 23 ft/sec/sec (0.72-g) under panic conditions from 80 mph, and with the present friction material fade is a moderate 17% in our 6-stop test. The handbrake, however, won't hold the car on a 30% grade—an unusual failing in a car with drum rear brakes.

Against the rather outdated character of the MGB is an impressive record of reliability, plus minimum servicing requirements. In short, it is an unfussy car, one well sorted out by several years of production. If driven with any degree of skill and cared for decently it is economical and trouble-free, and mileages of 70,000 before overhaul are common. Earlier irritations such as engine vibration, rear-axle noise and electrical problems are in the past now. By modern standards the B is not a "refined" motorcar, but one must spend considerably more than the B's $2947 to get refinement in a car of comparable character.

ALL-SYNCHRO MGB

SCALE: 10" DIVISIONS

PRICE

Basic list................$2810
As tested...............$2947

ENGINE

Type...............4 inline, ohv
Bore x stroke, mm.....80.3 x 89.0
  Equivalent in.......3.16 x 3.50
Displacement, cc/cu in.1798/110.5
Compression ratio.........8.8:1
Bhp @ rpm.........92 @ 5400
  Equivalent mph............99
Torque @ rpm, lb-ft...110 @ 3000
  Equivalent mph............62
Carburetion.........two SU HS4
Type fuel required......premium

DRIVE TRAIN

Clutch diameter, in..........8.0
Gear ratios: 4th (1.00).....3.91:1
  3rd (1.38)..............5.40:1
  2nd (2.17)..............8.47:1
  1st (3.44).............13.45:1
Synchromesh...........on all 4
Final drive ratio.........3.91:1
  Optional ratios...........none

CHASSIS & BODY

Body/frame: unit steel construction
Brake type: Girling: 10.8-in. disc
  front, 10.0 x 1.7-in. drum rear
Swept area, sq in...........310
Wheels.....wire spoke, 14 x 4½ J
Tires........Dunlop C 41 5.60-14
Steering type......rack & pinion
  Overall ratio............21.4:1
  Turns, lock-to-lock.........2:9
  Turning circle, ft.........32.0
Front suspension: unequal A-arms,
  coil springs, lever shocks, anti-
  roll bar
Rear suspension: live axle on multi-
  leaf springs, lever shocks

OPTIONAL EQUIPMENT

Included in "as tested" price: wire
  wheels, heater, tonneau cover,
  grille guard
Other: radio, folding top, overdrive,
  radial tires, chrome wire wheels,
  automatic transmission

ACCOMMODATION

Seating capacity, persons.......2
Seat width..............2 x 18.5
Seat back adjustment, deg......5
Driver comfort rating (scale of 100):
  Driver 69 in. tall..........80
  Driver 72 in. tall..........70
  Driver 75 in. tall..........65

INSTRUMENTATION

Instruments: 120-mph speedometer, 7000-rpm tachomete , fuel
  level, oil pressure, water temperature
Warning lights: alternator, brake
  fluid loss, high beam, directional
  signals

MAINTENANCE

Engine oil capacity, qt........5.0
  Change interval, mi.......6000
Filter change interval, mi.....6000
Chassis lube interval, mi.....3000
Tire pressures, psi.........21/24

MISCELLANEOUS

Body styles available: roadster as
  tested, coupe
Warranty period, mo/mi: 12/12,000

GENERAL

Curb weight, lb.........2220
Test weight.............2590
Weight distribution (with
  driver), front/rear, %....54/46
Wheelbase, in..........91.0
Track, front/rear.......49.2/49.2
Overall length..........153.2
  Width.................59.9
  Height................49.8
Frontal area, sq ft......16.6
Ground clearance, in......4.5
Overhang, front/rear...26.8/35.4
Usable trunk space, cu ft....2.9
Fuel tank capacity, gal......12.0

CALCULATED DATA

Lb/hp (test wt).............27.0
Mph/1000 rpm (4th gear)....17.6
Engine revs/mi (60 mph)....3410
Piston travel, ft/mi......1990
Rpm @ 2500 ft/min........4290
  Equivalent mph............77
Cu ft/ton mi................83.8
R&T wear index..............68
Brake swept area sq in/ton....239

ROAD TEST RESULTS

ACCELERATION

Time to distance, sec:
  0–100 ft................3.6
  0–250 ft................6.5
  0–500 ft...............10.0
  0–750 ft...............12.9
  0–1000 ft..............15.6
  0–1320 ft (¼ mi).........18.7
Speed at end of ¼ mi, mph....73
Time to speed, sec:
  0–30 mph................3.9
  0–40 mph................6.0
  0–50 mph................8.4
  0–60 mph...............12.1
  0–70 mph...............16.7
  0–80 mph...............23.2
  0–90 mph...............32.8
Passing exposure time, sec:
  To pass car going 50 mph....6.5

FUEL CONSUMPTION

Normal driving, mpg.......21–27
Cruising range, mi......250–325

SPEEDS IN GEARS

4th gear (5650 rpm), mph.....104
  3rd (6000)................79
  2nd (6000)................50
  1st (6000)................30

BRAKES

Panic stop from 80 mph:
  Deceleration, % g.........72
  Control............very good
Fade test: percent of increase in
  pedal effort required to maintain
  50%-g deceleration rate in six
  stops from 60 mph.........17
Parking: hold 30% grade......no
Overall brake rating........good

SPEEDOMETER ERROR

30 mph indicated.....actual 30.0
40 mph...................40.0
60 mph...................59.4
80 mph...................79.2
100 mph..................98.6
Odometer, 10.0 mi....actual 9.85

ACCELERATION & COASTING

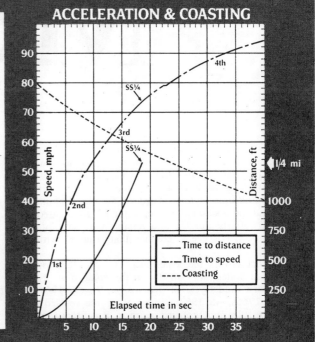

Time to distance
Time to speed
Coasting

Elapsed time in sec

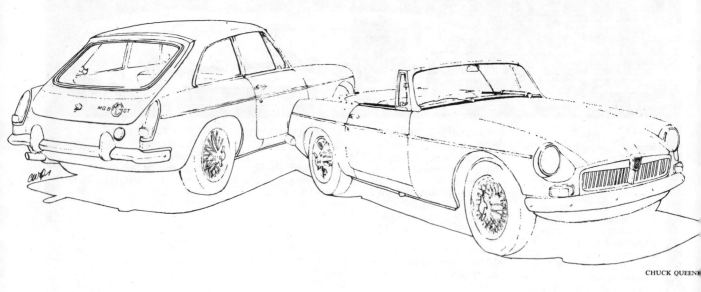

CHUCK QUEENE

Road & Track Owner Report:

MGB

OVER ITS FIVE years of production, the MGB has remained the second-best-selling sports car in America, outsold only by the domestic Corvette. And well it should; it offers good value for the money, lively performance and handsome styling. Word of mouth, personal experience and the occasional reader letter have indicated also that it was a fairly durable and reliable car, but not until now has it been possible to confirm that "rumor" with any authority.

This report covers the experience of 102 MGB owners in the United States and Canada. Of this total, 6 were from 1963, the first model year of production; 9 were 1964s, 14 1965s, 20 1966s, 45 1967s, and 8 were the revised 1968 with its new fully synchronized gearbox. Roadsters made up 74 of the total, the GT coupe (introduced in 1966) 28; 70 were bought new and the other 32 used. We eliminated from

the survey any new ones with less than 5000 miles on them or any used ones with less than 5000 miles by the present owners. The greatest concentration of miles on the odometer came between 10 and 20,000 miles, 30% of the cars having covered that range of miles. Nineteen percent of them were in their first 10,000, 14% between 30 and 40,000 miles, 13% between 20 and 30,000, 10% between 50 and 60,000. The highest mileage reported was 83,000.

Since the MGB is a sports car we were a little surprised at the owners' reasons for buying one: number one reason was styling, given by 52%. Next came handling, with 24%; price, 21%; reliability, 17%. Eighteen percent were less articulate, giving reasons such as "reputation," magazine road tests and word-of-mouth as their reasons for purchasing a B. As expected, nearly half of the GT owners specified that they were looking for a "sports car with GT comfort."

24

Previous ownership, not anticipated as a frequent reason for buying a popular sports car, was given by 9% of the owners —reflecting that MG is, after all, the pioneer in the American sports car market. The availability of parts and service, as compared to similar cars' dealer networks, was given by 6%, as were seating comfort and fuel economy. Some of the miscellaneous responses to the question, "Why did you choose MGB?" further explain the car's character: "fun," "easy to work on," "maneuverability" and so on.

MGB owners fall right into the national average when it comes to the amount of driving they do—43% put 10-15,000 miles per year on their Bs. Thirty-four percent drive 15-25,000 miles each year, 4% over 25,000; nobody reported less than 5000 per year. And, not at all surprising, 99% of them use the car for daily transportation—that's what it's intended for. Only one of the owners reporting said he raced his car. However, fully 35% of them do participate in rallies and 14% in slaloms. Fifty-seven percent depend upon their Bs for long trips and most of them consider the car satisfactory for extended mileages. The bulk of B owners —53%—drive their cars "hard"; 42% said they drive them "moderately" and the rest drive "very hard." Searching for some correlation between hard driving and mechanical trouble, we examined the cars in which 30,000 or more miles had been covered and found that those who drove moderately had a one-in-six (16%) chance of some major mechanical work by 50,000 miles whereas those who drove hard or very hard had more than a 50% chance of same. The most commonly mentioned trouble among those who drove vigorously was gearbox synchronizers wearing out—at mileages as low as 24,000 but generally more like 45,000.

Asked if they follow the manufacturer's preventive maintenance and lubrication recommendations, a solid 65% said they did, 23% said that they mostly followed the schedule and 12% said flatly that they didn't bother. Fully 14% change oil more often than the 3000-mile (early) or 6000-mile (later) recommendations—many enthusiasts just don't trust the decade's progress in oil and the filtering thereof! Eight percent change the filter more often than recommended, 5% lubricate their chassis more often. Not surprising is that 9% do all their own maintenance.

How about the MGB owners' opinion of their dealers? The dealers didn't fare very well in our survey. Only 39% (all owners had tried dealers at least once) rated "good" in the B owners' eyes, and 34% "fair." And 27% were called "poor." Owners commented on the high service costs in 12% of the questionnaires. Using Volkswagen as a base-line for comparison, of the questionnaires tabulated so far on that make, 62% considered their dealer "good," 26% were ranked as "fair" and 12% as "poor," with 9% commenting on high service costs. As this series progresses it will be interesting to see if owners of other makes have as low an opinion of their dealers as MGB owners.

We asked the B owners what they liked best about their machines. Resoundingly, handling was the No. 1 quality—named by 58% of them. Next came fuel economy (25%), performance (23%), reliability and durability (25%), styling (22%), and driving fun (12%). Comfort was highly rated for a fairly small sports car, being named by 12% of the roadster owners and 33% of the GT owners, and legroom was appreciated by 9%. Miscellaneous favorite features were the brakes, shift linkage, the optional overdrive, the instrument layout and the GT's "tight feel."

And what did they like least about their MGBs? Clearly again, the old gearbox with its noisy, unsynchronized first gear (the 1968 model has a new gearbox that corrects both objections). The roadster top was resented by 17% and another 13% named "weatherproofing" without being specific. Ventilation was a sore point with 13% and, related to it, extreme heat around the gearbox tunnel bothered 10%. Also, 9% specified the heater among their biggest gripes. Material quality was criticized by 8%, with specific reference to the plastic interior door and window handles. Other, less frequent grouses were aroused by parking damage vulnerability, hard steering and rear end noise (the last of which has been corrected in production).

The instruments, although appreciated by owners as legible and handsome, are also the biggest reliability deficiency in the MGB: 22% of the owners had one or more—very often more—instrument failures. Only the fuel gauge escaped—tachometer, speedometer, odometer, water temperature and oil pressure gauges failed with remarkable regularity. One 1967 owner had six of the (new in 1966) electric tachs fail during his warranty period!

Next on the list of common failures was the exhaust system. Everyone seems to like the B's mellow exhaust note, but the bonded rubber hangers on which it rides are obviously failure-prone—21% of the owners have had repeated failures often with broken mufflers, resonators or headpipes resulting.

The cooling system also caused trouble in 21% of the cars surveyed. "Overheating" was quoted by most of these though there was also an indication that radiator hoses can be expected to fail at about 30,000 miles.

Oil leaks—from engine, gearbox and differential—were experienced by 12% of our owners. Likewise, 12% had their gearbox synchronizers wear out—but this appears to be directly attributable to hard driving, as explained above. Twelve percent of our owners experienced the dreaded silence of fuel pump failure. And 11% had undue brake trouble of some kind—premature wear, loss of braking in wet weather and loose disc pads were named. Starter failure was reported by 7%, the finger of blame pointing to the position and lack of shielding which make it vulnerable to corrosion. Paint fading and premature tire wear were reported each by 7%, most of the latter occurring on the heavier GTs. General electrical problems cropped up in 6% of the cars, but these became more common than that— more like 20%—when the mileage gets past 40,000. Nothing else rates mention on the trouble list except the famous running-on (MG owners seem to accept it) and poor starting below 0°F—the latter not reported often because the reader sample was weighted toward warmer climates.

The final—and perhaps the most telling—question is "Would you buy another of the same make?" Seventy percent said that they would; 19% said absolutely not, 11% weren't sure. Half of those who "will not" intend to go to more sophisticated and expensive cars, like Porsches, stating that they are satisfied with their Bs as far as they go. Bad service, outdated engineering and unreliability were quoted by those not intending to buy MGs again.

Generalities from the questionnaires that may be of interest include the fact that tire life averages about 30,000 miles on the roadster, somewhat less on the GT (because of its greater weight) and that 9% of the owners had re-shod their Bs with radial-ply tires. It is also apparent that front disc brakes need new pads before the rear drums need new linings and that owners should begin to check pad wear at about 15,000, though most will get more miles than that.

Among those whose Bs required major work, it appears that gearbox overhauls occurred at about 45,000 miles but really need not to have happened at all. Clutches seem to have about the same longevity—and to fail for the same reasons. Only two engine overhauls were reported by the readers surveyed and both happened at about 50,000 miles. Based on the experience of other owners—as well as personal experience—we would say that 70,000 is the mean engine life between overhauls. Our highest mileage owner had gone 83,000 without even a valve job. It also seems that the differential will last the life of the car unless severely abused and that only those who corner hard habitually will need front end work before 50,000 miles.

Of the survey results that might require comment, we wonder whether some of the "overheating" reported might have resulted from owners interpreting 200°F as overheating. These days, in summer weather, 200°F is to be expected in most cars as a matter of course. With the 7-psi pressure cap of the MGB, the boiling point is over 230°F at sea level. We simply cannot resist commenting on the number of instrument failures reported and would like to point out that this is one area where driving habits cannot be blamed for failure. As such a great proportion of these are replaced under warranty we cannot help but wonder about the wisdom of the economics that does not require better ones in the first place.

A few direct quotes from the owners: "I always feel that I am in full command of the car." "That damn top takes an engineering degree to put down!" "Fuel pump replaced. I consider this normal, however, as this is the third MG I've owned." "The steering is very hard, a bit too much for my wife, but it's sure and it goes where it's pointed." "High torque makes downshifting optional in almost all driving situations." "Fuel economy surprised me, at 27 mpg, with effortless driving at 70-80 mph even in the high land of Colorado and Utah."

That's the MGB, then; simple and durable, economical of fuel, fun to drive, short on weatherproofing and with a spotty reliability record so far as some components are concerned. Its owners are a fairly happy group, however, and in the long run this counts for a lot.

CONTINUED FROM PAGE 20

the city I get a shade over 20 mpg, open roads and turnpikes always have yielded over 25, and for the total period of ownership mileage has averaged 24.2 mpg. Needless to say gasoline has not been the major expense. Insurance gets that dubious honor. Single males under 25 can expect to pay around $425 a year in suburban Southern California, including comprehensive and collision. Living in Chicago or New York will add another $300 on top of that if full coverage is to be retained. If you're older, it's a different proposition entirely. Dealer maintenance is also more expensive in the Midwest and East than in California. The car has been maintained by authorized dealers at the specified intervals.

General driving impressions can be obtained from published road tests but a few additional points should be made. The tonneau cover and top have always fit quite well. But the thread in the seams was pretty much disintegrated after 18 months; easily remedied if you know a girl who can sew. The top is not completely airtight but no snow got into the driving compartment during this winter's super storm, which left the car buried in a 6-ft snow bank. The car has surprisingly good traction in snow and ice, my only complaint about winter driving being that it tends to bottom out in the ruts and grooves of unplowed residential streets. (Thank *you*, Mayor Daley.) Tuned for winter weather and using a 190° thermostat, the car starts easily and the heater provides enough heat after everything is warmed up. The car is also fairly comfortable for cross-country driving, two drivers making 800 miles a day without strain. Two general complaints are the stiff throttle, which drivers new to the car find difficult to manage, and windshield wipers that don't clean well and do it at one speed only.

Without a doubt the biggest annoyance with the car is the exhaust system, a problem that shows up at about 5000 miles and each 5000 thereafter. The original muffler was poorly mounted, and brackets and welds were always breaking. American workmanship doesn't seem to be much better and it takes two hands to count the number of times my muffler has fallen off. Four different systems have been used to date and a few new resonances and rattles are cropping up again. Maybe one of those expensive Abarth systems wouldn't be so expensive.

The other major difficulties during the car's history have been the recent change of life, which has proven both aggravating and expensive. A complete electrical shutdown on a freeway in New York was caused by a worn-out generator and dead batteries. Those batteries in the back are a nuisance in general as any MG owner will attest to. Fuel starvation in the middle of the New Mexico highlands was traced to holes in the pumping diaphragm. ("Don't see too many of these cars here in Tucumcari!") Fuel is now pumped by an AC light truck unit. Jets sticking in the carbs required a rebuild on that part of the fuel feed system. Loud clanks from the drive train were remedied by new U-joints. The rest of the engine seems to be in fair shape, except that one of the exhaust valves is burned. A phenomenal rate of oil burning was traced to the crankcase vent.

Tires have also been somewhat of a nuisance and expense. The original Dunlops lasted about 14,000 miles and were replaced by Pirelli tread recaps from a large Southern California capper. Two of these were bad. One was capped cockeyed, and the blame was mislaid on my wire wheels. "Those wheels must be bent; we don't turn out out-of-round caps." My foot. The other bad cap started to disintegrate after about 5000 miles, which was enough to put me many miles away from their place of purchase. I'm not favorably impressed by recaps. The Cinturato is a great tire—that I have had some bad luck with. Two flats at freeway speeds racked out one carcass, and two flats in the back yard were attributed to pinched tubes. The radial-ply tire does increase the gas milage considerably, even after correcting the odometer 5% because of the lower profile. Put another way, one trip from Brooklyn to Chicago on the new Pirellis was made in 14 hours on 29 gallons of premium, and that's 860 true miles—figure it out. I was stunned too, but it has kept on doing it. But oh, for an overdrive so I could cruise at 3300 rpm instead of 4000-plus.

I was somewhat taken aback when I sat down and added up all the expenses. Now I know why I'm always broke. At 11¢ a mile it doesn't really strike me as being cheap, and during the last year it hasn't been all that reliable. But I must admit it was fun, especially back in California, with all those nice canyons and sunny days. It's even fun here in Chicago, especially when a Checker Cab driver leans out his window and says, "Hey, man, is that one of those Jaagwaars?" Yes, there are places where sports cars are somewhat a rarity. Would I buy another one? Not now. When running, it is almost as quick as a new one, it is all paid for, and it is still an appealing car, and it's cheaper to replace components than to buy a whole new car—I'll probably keep it for another 45,000 miles.

MGC

Engine swap turns the MGB into a 6-cylinder smoothie

WE MAY discourage our readers from engine swaps, but the British motor manufacturers certainly don't balk at them. When the classic Austin-Healey vanished from the market, partly because it would have been uneconomic to try to make it conform to U.S. safety regulations and partly because it would have been uneconomic to supersede it with a new Healey using its own unique body and chassis, BMC (now part of the combine BLMC) decreed that it should be "replaced" by an MGB with the big Healey 6-cyl engine.

Actually, it isn't as simple as this; engine swaps never are. To start with, the engine itself was a venerable design, in need of either total redesign or at least updating. It got the latter: a new crankcase with less overall length, re-spaced cylinder bores and seven main bearings instead of four. This bit of redesign allowed a small weight reduction from 611 lb to 567 (with clutch)—but when we compare this with, say, the 5-liter Ford V-8 at about 480 lb it is obvious that an altogether new engine could have been much lighter. Using slightly smaller (HS6 instead of HS8) carburetors than the Healey, the revised 3-liter engine produces 145 bhp at the same 5250 rpm at which its predecessor developed 150 bhp.

Fitting the big "six" into the space occupied by a 1.8-liter four without lengthening the car required extensive alterations. A larger transmission hump was required to accommodate the optional Borg-Warner automatic transmission and the two big longitudinal frame members are spread farther apart to more rigidly accept a gearbox support crossmember that will be loaded by over 50% more engine torque.

27

The front crossmember that completely carries the coil-spring front suspension in the MGB had to be considerably trimmed to allow the longer engine to pass over it, leaving insufficient structure to do this whole job in the C. The solution was to design new lateral links replacing the B's unequal-length A-arms and attaching to new pivot points (the static roll center, at ground level, is unchanged) and to do the springing with torsion bars attached to the lower links and anchored at the rear to the same crossmember that supports the gearbox. It should be emphasized here that torsion bars were selected because they were a solution to the space problem—not because they possess any significant springing properties that coil springs do not. A 0.687-in. front anti-roll bar is standard.

Rear suspension is the same as for the B except that the springs have seven leaves instead of five. Spring rates measured at the wheels are 100 lb/in. front and rear (vs. 73/93 for the B), giving natural frequencies of 74 cycles/min front and 87 cycles/min rear. Frequencies for the B—and these are the real criterion of ride firmness—are 69 front, 88 rear. Thus the C is slightly stiffer in the front than the B and at the same time more evenly matched front-to-rear so that it should be less prone to pitch. Weight distribution is less different from the B than we expected: 55% front, 45% rear vs 54/46 for the B; overall weight is up from 2220 lb to 2600.

The front bulkhead carrying the engine radiator has been moved forward, using the B's considerable empty space up there. The steering rack also had to be moved, and though early MGCs had the same steering ratio as Bs the current ones have a 24.0:1 overall ratio vs. the B's 21.4. Other changes are in the drive train: the new all-synchro gearbox and final drive ratios more appropriate to a large-displacement engine. MGCs sold in England have a 3.07:1 final drive or 3.31 with the optional overdrive; for the U.S. we get the 3.31 without overdrive and a 3.70 with. Finally, there is a change in wheels and tires, from 5.60-14 with 4½-in. rims on the B to 5.60-15 with 5-in. rims on the C.

Trim, body and interior details are little different from the B. A clumsily handled hood bulge, with another small pod on top of it to clear the carburetors, accommodates the 6-cyl engine, and the C becomes the first MG roadster to offer readily adjustable seatbacks.

Driving the MGC

THE FIRST sound heard upon starting the MGC is a soft burble from the exhaust system quite like the B's pleasant note. As the revs increase the note becomes more distinctly "six"-ish though remaining mellow. Underhood, the engine is very quiet, as was the Healey, and noticeably smoother—but a noisy cooling fan is prominent. Strangely enough, the engine seems torqueless at low speed and subjectively lacks the punch of the old Healey, but when we compared the acceleration times we got with the C to those of the last Healey we tested, we found the C only slightly less quick, a difference easily explained by the smaller carburetors and taller gearing of the C—which weighs only some 50 lb less than that Healey. Perhaps it's the extra smoothness and quietness of the car that make the performance seem less exhilarating, but more likely it's that four years later we'd expect more performance from an updated model, not less.

The engine is certainly flexible enough in everyday use, though it's a reluctant starter from cold. We found no side effects from its air-injection emission control other than the familiar lag in returning to idle. Fuel economy isn't especially good at 17.8 mpg.

MGC
AT A GLANCE

Price as tested............................$3637
Engine................6-cyl, ohv, 2912 cc, 145 bhp
Curb weight, lb.............................2600
Top speed, mph..............................118
Acceleration, 0–¼ mi, sec....................17.7
Average fuel consumption, mpg................17.8
Summary: not as nimble but quieter & faster than MGB . . . dated body & suspension . . . decent value for money.

GORDON CHITTENDEN PHOTOS

ROAD TEST
MGC

SCALE: 10" DIVISIONS

PRICE

Basic list.................$3410
As tested................$3637

ENGINE

Type...........6 cyl inline, ohv
Bore x stroke, mm....83.3 x 88.9
  Equivalent in.....3.28 x 3.50
Displacement, cc/cu in..2912/178
Compression ratio..........9.0:1
Bhp @ rpm.........145 @ 5250
  Equivalent mph (OD)......132
Torque @ rpm, lb-ft..170 @ 3500
  Equivalent mph (OD).......87
Carburetion.........two SU HS6
Type fuel required.....premium
Emission control......air injection

DRIVE TRAIN

Clutch diameter, in..........9.0
Gear ratios: OD (0.820)....3.03:1
  4th (1.00)............3.70:1
  3rd (1.31)............4.84:1
  2nd (2.06)............7.61:1
  1st (2.99)...........11.03:1
Final drive ratio..........3.70:1

CHASSIS & BODY

Body/frame unit steel construction,
  aluminum hood
Brake type: 11.25-in. disc front,
  9.0 x 2.5-in. drum rear
  Swept area, sq in.........353
Wheels....chrome wire knockoff,
  15 x 5
Tires........Dunlop SP41 165-15
Steering type......rack & pinion
  Overall ratio.........24.0:1
  Turns, lock-to-lock........3.5
  Turning circle, ft........34.5
Front suspension: unequal-length
  A-arms, torsion bars, tube shocks,
  anti-roll bar
Rear suspension: live axle on multi-
  leaf springs, lever shocks

EQUIPMENT

Options on test car: overdrive
  ($175, see text), chrome wire
  wheels ($150), radial tires ($40),
  tonneau cover ($37)
Other: automatic transmission
  ($255), hardtop ($230)

ACCOMMODATION

Seating capacity, persons.......2
Seat width.............2 x 18.0
Head room.................38.8
Seat back adjustment, deg....45
Driver comfort rating (scale of 100):
  Driver 69 in. tall...........85
  Driver 72 in. tall...........75
  Driver 75 in. tall...........70

INSTRUMENTATION

Instruments: 140-mph speedo,
7000-rpm tach, 99,999 odo, 999.9
trip odo, oil press, water temp,
fuel level
Warning lights: alternator, brake
fluid loss, directionals, high
beam

MAINTENANCE

Engine oil capacity, qt.........7.2
Every 3000 mi: lube chassis, fill
  carb dampers, var. op'l checks
Every 6000 mi: chk valve clear-
  ances, ign. system; cln plugs,
  chg engine oil & filter
Every 12,000 mi: chg air filters,
  tune engine, chg plugs
Every 24,000 mi: chk OD filters
Tire pressures, psi.........26/22

GENERAL

Curb weight, lb............2600
Test weight...............2915
Weight distribution (with
  driver), front/rear, %....55/45
Wheelbase, in.............91.0
Track, front/rear.....50.0/49.2
Overall length............153.2
  Width................59.9
  Height................49.8
Ground clearance, in.........4.4
Overhang, front/rear...26.8/35.4
Usable trunk space, cu ft......2.9
Fuel tank capacity, gal.......14.0

CALCULATED DATA

Lb/hp (test wt)............20.1
Mph/1000 rpm (O'drive).....24.7
Engine revs/mi (60 mph)....2430
Engine speed @ 70 mph, rpm.2830
Piston travel, ft/mi.........1422
Cu ft/ton mi (4th gear).....105.0
R&T wear index (O'drive)....34.6
R&T steering index.........1.208
Brake swept area sq in/ton....242

MISCELLANEOUS

Body styles available: roadster as
  tested, GT coupe
Warranty, mo/mi......12/12,000

ROAD TEST RESULTS

ACCELERATION

Time to distance, sec:
  0–100 ft................3.5
  0–250 ft................6.4
  0–500 ft................9.7
  0–750 ft...............12.5
  0–1000 ft..............15.1
  0–1320 ft (¼ mi).......17.7
Speed at end of ¼ mi, mph....80
Time to speed, sec:
  0–30 mph...............3.7
  0–40 mph...............5.4
  0–50 mph...............7.2
  0–60 mph..............10.1
  0–70 mph..............13.6
  0–80 mph..............17.8
  0–100 mph.............32.6
Passing exposure time, sec:
  To pass car going 50 mph....7.5

FUEL CONSUMPTION

Normal driving, mpg.......17.8
Cruising range, mi..........249

SPEEDS IN GEARS

O'drive (4780 rpm), mph......118
4th (5600).................115
3rd (5600).................86
2nd (5600).................54
1st (5600).................39

BRAKES

Panic stop from 80 mph:
  Deceleration, % g..........84
  Control............very good
Fade test: percent of increase in
  pedal effort required to maintain
  50%-g deceleration rate in six
  stops from 60 mph.........51
Parking: hold 30% grade......no
Overall brake rating........good

SPEEDOMETER ERROR

30 mph indicated.....actual 29.9
40 mph.....................39.5
60 mph.....................58.2
80 mph.....................77.2
100 mph....................96.0
Odometer, 10.0 mi.....actual 9.67

ACCELERATION & COASTING

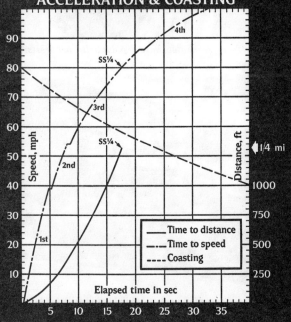

MGC

The all-synchro gearbox, which we had already tried on the revised B, is a most satisfactory unit, up to modern standards in every way, and the synchronized 1st gear means it's possible to come up with a more useful spread of ratios. British Motor Car Distributors, the local distributor who loaned us the test car, is ordering all Cs with overdrive ($175 extra), which is nice but not so essential as we think it is in the B. The test car had a clearly audible final drive, even over the considerable wind roar of the roadster top, but not to what we consider an objectionable level.

The C rides about like the B—choppy, abrupt movements over sharp road irregularities but fairly soft on good roads. Its increased weight, not compensated by increased spring travel, means that it's more prone to bottoming than the B and one has to be careful about dips. The body structure is exceptionally rigid and rattle-free, but the limited travel and relatively archaic rear suspension don't encourage fast travel on indifferent roads.

In spite of the increased steering ratio and a huge steering wheel, steering efforts have gone up from an already high level in the B; still, the C doesn't quite seem a candidate for power steering. On the recommended tire pressures of 26 psi front, 22 psi rear there is very little chance of power oversteer except at low speeds, though we found these pressures to be less satisfactory on wet pavement than equal inflation all around. The C gets around a given smooth corner at a respectable speed if the driver uses the slight amount of off-throttle oversteer available; final oversteer that occurs without any throttle-jockeying seems less part of MGC handling than in previous MGs. Overall, we would rate the C's handling as decent but not particularly entertaining.

One sits low in the cockpit and the new steering wheel seems positively huge, often brushing the thighs of the driver. Seats are the primary deviation from the B, having adjustable seatbacks, but they are of traditional MG design—that is to say, rather soft and more like chairs than many current automotive seats.

To meet the 1969 Federal regulations about wiped area of windshields, MG has gone to a Jaguar-like array of three wipers; these really do the job, though the left one on our car didn't park correctly. But outward vision isn't a strong point of the B-C body, the windshield and window tops being quite low and the rearview mirror forming a large blind spot. Our test car had a funny—unattractive but functional—combination of outside mirrors, one on the left door and one on the right front fender. Items like this and the unattractive safety dash (which is laid *over* the earlier dash that is still used on domestic MGs and which does away with the glove box) indicate that BMC is inclined to take the cheapest and least imaginative approach to the demands of safety legislation.

Interior, trunk and exterior body details are all identical to those of the B except the seats as noted above. The body shows much evidence of being long in the tooth, now being rather old-fashioned in appearance as well as in details like the hood and deck-lid props and that terrible roadster top that has to be taken off and stored in the trunk.

The MGC is still a small roadster, but it's not nimble enough to really make its mark as a sports car; rather, it scores as a personal long-distance cruiser. If the record of the old Austin-Healey engine applies, the engine should be nearly unbreakable, and the GT coupe version (at $400 more than the roadster) should be especially satisfactory for open-road use. We would suggest that the car might be sharpened up for the American market with wider wheels (possibly cast alloy ones instead of wires) and styling touches that make it more attractive than the B instead of less (ouch, that hood bulge). However, this roadster, priced at $345 over a similarly equipped B, is a reasonable combination at the price and makes up what it lacks in refinement (when compared with its competition) with a liberal dose of performance of the kind that small-displacement fours can't deliver.

Safety-ized cockpit has ugly thick padding on passenger side.

Seats are part-way reclinable, have 3-point belts, headrests.

Big 6-cyl engine occupies most of available space under hood.

FOUR SPORTS CARS

How did the Fiat 124 Spider, MGB Mk II, Porsche 914 & Triumph TR-6
measure up on a 1250-mile run?

R&T Comparison Test

IT WOULD BE POSSIBLE for a regular reader of *Road & Track*, interested in buying an open sports car in the $3000-4000 class, to rummage through his back issues for the appropriate road tests. He'd find prices, specifications and performance data, plus objective and subjective comment on the four cars which qualify: the Fiat 124 Spider, MGB, Porsche 914 and Triumph TR-6. In effect, this comparison test gathers together, updates and arranges all that material in convenient tables; more important, by driving

the four cars under identical conditions a combined total of 5000 miles, we've been able to make an accurate assessment of the practical, functional and aesthetic qualities of each and to arrive at an overall rating of preference.

Except for the fact that they are open sports cars within the specified price range, the four are surprisingly dissimilar. The Triumph has a generous 2.5-liter 6-cyl engine, while the three smaller 4-cyl units vary considerably: the Fiat makes up in part for its small displacement by having a modern belt-driven dohc system, the MG's older ohv engine has less peak horsepower but far greater torque, and the Porsche has a VW-built aircooled flat-4. In chassis design, the Fiat and MG have unit structures with live rear axles, the Triumph has a separate body and frame with independent rear suspension of limited effectiveness, and the Porsche is unique in having a mid-positioned engine, naturally with independent rear suspension. To transmit power the Triumph and MG, with the largest displacements and greatest torque, have 4-speed gearboxes, while the smaller-capacity Fiat and Porsche employ 5-speeds (even here notably different in design). In body configuration, the Fiat has a recent design of rather

BASIC SPECIFICATIONS: 4 SPORTS CARS

	Fiat 124 Spider	MGB Mk II	Porsche 914	Triumph TR-6
Curb weight, lb	2090	2220	2085	2360
Test weight, lb	2420	2590	2455	2680
Distribution f/r, %	55/45	54/46	46/54	51/49
Wheelbase, in	89.8	91.0	96.4	88.0
Track, f/r	53.0/51.8	49.0/49.3	52.6/54.1	50.2/49.8
Length	156.3	152.7	156.9	156.0
Width	63.5	59.9	65.0	58.0
Height	49.2	49.4	48.0	50.0

FOUR SPORTS CARS

classic proportions and the only +2 seating in the group, the MG and Triumph are both very narrow with dated lines, and the Porsche is again unique in having a low nose, retractable headlights, an integral rollbar with removable roof panel and two luggage compartments. The greatest similarities among the four cars—besides their common purpose of providing open-air driving pleasure—are their weights and dimensions.

A test crew of six (including Editor-at-Large Henry Man-

ney, whose irreverence tempered the more routine approach of the others) took turns behind the steering wheels and in the passenger seats during the 3-day, 1250-mi run from Newport Beach to the Grand Canyon and back. The experience included altitudes from near-sea level to over 7000 feet; terrain from flat, featureless desert highway to narrow mountain hairpin turns; average speeds from 45 to 85 mph; and such misfortunes as two flat tires on the MG (from staples picked up during the skid-pad cornering test), locking ourselves out of the Triumph (solved without damage by a straightened coat hanger), electrical accessory maladies on the Fiat (the result of fording an axle-deep creek), and a still-unexplained engine failure in the Porsche (which quit

ENGINE & DRIVE TRAIN: 4 SPORTS CARS				
	Fiat 124 Spider	MGB Mk II	Porsche 914	Triumph TR-6
Engine type	dohc 4	ohv 4	ohv flat-4	ohv 6
Bore x stroke, mm	80.0 x 71.5	80.3 x 89.0	90.0 x 66.0	74.7 x 95.0
Displacement, cc	1438	1798	1679	2498
Bhp @ rpm	96 @ 6500	92 @ 5400	85 @ 4900	104 @ 4500
Torque @ rpm	82.5 @ 4000	110 @ 3000	103 @ 2800	142 @ 3000
Transmission	5-speed	4-speed	5-speed	4-speed
Final drive ratio	4.10:1	3.91:1	4.43:1	3.70:1
Engine speed @ 70 mph	4000	3980	3120	3370

Fiat interior: wide, neat, comfortable.

Porsche's: modern, simple, no nonsense.

without warning while cruising at 70 mph on the desert, showed no signs of faulty electrics or vapor lock, started up again without fuss and gave no further trouble). With the exception of finding a tire repair service that would work with wire wheels on a Sunday morning in Kingman, Arizona, none of these incidents required more than five minutes' concern.

Every car performed satisfactorily on the trip, but in switching back and forth from car to car, we had established clear preferences at the end of it. Our score sheets compared the cars in 14 categories, with ratings of 1 (best) to 4 (worst) in each: engine, gearbox, steering, roadholding, ride, brakes, structural integrity, seating, interior fittings, exterior appearance, weather protection, heating/ventilation, carrying capacity and accessibility/maintenance, plus a separate rating of overall preference. Totalling the 14 categories, the Fiat scored 48 "bests", the Porsche was next with 22, the Triumph third with 13 and the MG last with only one. The Fiat also got the least number of "worsts" with 6, the Triumph was next with 15, the Porsche third with 23 and the MG again last with 40. In the overall rating of preference the Fiat received five "best" votes and the Triumph one; the Porsche got two "worsts" and the MG four. So the Fiat was clearly the best overall car in the opinion of the test crew and the MG the worst. What about the Porsche and Triumph? The uncompromising Porsche, perhaps in a reflection

Triumph's: narrow but handsomely detailed.

MG's: snug with big wheel, dash padding.

GENERAL DATA: 4 SPORTS CARS

	Fiat 124 Spider	MGB	Porsche 914	Triumph TR6
Basic list price	$3300	$3150	$3695	$3595
Price as tested	$3480	$3360	$3895	$3810

Asterisked prices include: for Fiat, AM/FM radio; for MG, tonneau cover, radial tires, delivery and handling; for Porsche, appearance group, AM/FM radio, local transportation, pre-delivery; for Triumph, tonneau cover, AM/FM radio. *Includes dealer preparation. †Preparation charge added by dealer.

	Fiat 124 Spider	MGB	Porsche 914	Triumph TR6
Chassis type	unit	unit	unit	separate
Brake type	disc/drum	disc/drum	disc/disc	disc/drum
Swept area, sq in./ton	245	259	383	248
Suspension type, f/r	ind/live	ind/live	ind/ind	ind/ind
Standard tires	165-13 rad	5.60-14 bias	165-15 rad	165-15 rad
Drivers' Comfort Rating:				
For driver 69-in. tall	85	80	85	
For driver 72-in. tall	80	70	65	75
For driver 75-in. tall	75	65	60	75
Steering, lock-to-lock, turns	2.75	2.9	2.5	3.25
Steering Index	0.938	0.928	0.902	1.073
Usable trunk space, cu ft	6.2	2.9	9.9	4.3
Fuel tank capacity, gal	11.8	12.0	16.4	13.5

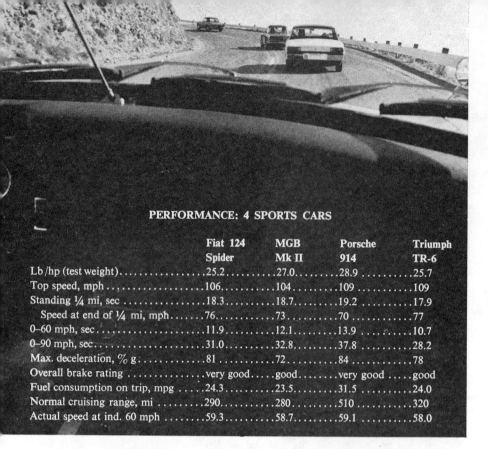

PERFORMANCE: 4 SPORTS CARS

	Fiat 124 Spider	MGB Mk II	Porsche 914	Triumph TR-6
Lb/hp (test weight)	25.2	27.0	28.9	25.7
Top speed, mph	106	104	109	109
Standing ¼ mi, sec	18.3	18.7	19.2	17.9
Speed at end of ¼ mi, mph	76	73	70	77
0–60 mph, sec	11.9	12.1	13.9	10.7
0–90 mph, sec	31.0	32.8	37.8	28.2
Max. deceleration, % g	81	72	84	78
Overall brake rating	very good	good	very good	good
Fuel consumption on trip, mpg	24.3	23.5	31.5	24.0
Normal cruising range, mi	290	280	510	320
Actual speed at ind. 60 mph	59.3	58.7	59.1	58.0

Taking down and putting away MG's many-piece top is a monumental exercise.

Lowering Triumph's top involves some folding but is easier than it looks.

Stowing Porsche's roof in trunk is easy but Fiat's top is model of simplicity.

FOUR SPORTS CARS

of it uniqueness, elicited more extreme reactions (more "bests" *and* "worsts") but in the final tabulation just edged out the Triumph for second:

1	Fiat 124 Spider	139 points (low score best)
2	Porsche 914	207
3	Triumph TR-6	213
4	MGB Mk II	279

Fiat 124 Spider

THE OVERWHELMING MARGIN of preference for the Fiat surprised everyone, including those who voted most strongly for it. The Fiat won by virtue of its overall balance. On the road its steering, roadholding and ride made an unbeatable combination; in pampering the occupants its seating, interior fittings and weather protection were most effective, approached only by the Porsche.

Every car in the group has a serious flaw; the Fiat's is its small, low-torque engine. As the performance table shows, this fast-turning dohc unit was second only to the much bigger Triumph in through-the-gears acceleration, but on the uphill sections of the mountain roads we continually had to use the lower gears and listen to unpleasant engine racket to keep up. For a 5-speed, the gearbox is well designed though we were able to beat the synchromesh frequently. The basic understeer and low-speed steering heaviness were not appreciated at first; after miles of fast driving, the Fiat's excellent transient cornering characteristics and good absolute adhesion (a close second to the Porsche on the skid pad) made it the most enjoyable car. In steady high-speed cruising the Fiat surprised us by its low engine noise (despite the high engine speeds); it also had the least wind noise.

The Fiat's top was the best by far, giving almost closed-car protection when up and requiring only 10 sec of one man's time to put up or down: a model of functional design. If we were to characterize the Fiat with two words, we'd say **Modern** and **Civilized.** To keep its favorable position in the sports car market, the 124 Spider (and its GT sister, the Coupe) will need the promised 1608-cc engine. With 20 percent more power and 25 percent more torque, the car will have the mid-range performance which the rest of its modern specification cries for. Maybe in 1971 or even late 1970?

Porsche 914

As WE'VE SAID, the 914's very unconventionality is responsible for both the high and low marks it received. Every feature was either praised or strongly disliked. Its awkward styling, noisy VW engine and cantankerous shift pattern put most drivers off at the start, but its excellent roadability, comfort and long-legged cruising qualities (both 4th and 5th ratios are below unity) found favor.

Like the Fiat, the Porsche does not pull well in the upper gears; in this case it is the tall gearing rather than lack of torque which is to blame. This would be less objectionable if the gears were easier to select; only our most experienced road tester found the gearbox acceptable after getting used to it. The straight-line performance of the Porsche is the lowest of the four though the 0.708:1 top gear gives it a maximum speed equal to the Triumph's. Its brakes are the best.

Though the styling of the Porsche (particularly the front) was unpopular, the body was found excellent in terms of function and convenience. The integral rollbar and lift-off, easily-storable roof panel have to be *the* design for the future, whether safety regulations demand it or not. The interior of the 914 is extremely handsome and, like the Fiat's, roomy. The mid-positioning of the engine gives the benefit of an extra trunk but hardly improves the handling (compared to a 911); this may be better as the car is developed. The 914's excellent fuel economy and large tank give an unprecedented cruising range of 500 miles!

In two words, the Porsche 914 is **Futuristic** but **Unresolved.** It has the ingredients of a fine sports car, but at present is not sophisticated enough for its high price. A handsomer nose, better shift linkage and minor suspension development would do the job.

Triumph TR-6

THE STRONG, BEAUTIFUL-SOUNDING engine of the TR-6 makes it one of the easiest and most enjoyable cars to drive—on good roads—and on this basis alone it almost scored above the Porsche overall. Because of emission requirements, peak power of the 2.5-liter six is down to a moderate (for the group) 104 bhp but torque is so healthy

that the Triumph can move away from the other three cars at will and has the added advantage of relaxed gearing. Fuel consumption, seemingly a function of power output, is as low as for the smaller-displacement MG and Fiat. Roadholding and ride are fine until you encounter rough surfaces—then the car's extremely dated chassis and suspension make themselves known all too harshly.

The dated character of the car also shows up in the styling and narrow cockpit, but in each case Triumph has done a better than fair job of improving a design that goes back to the TR-4 of late 1961. The ends of the TR-6 are quite successfully modernized and the interior is very handsome and well detailed (especially the dashboard), but nothing short of an all-new structure can cure the narrowness (not helped by the short wheelbase which forces the long engine to intrude on the occupants). The top is nowhere near as convenient to put up or down as on the Fiat or Porsche, requiring a lot of careful folding, but is still vastly better than the disassembled-reassemble affair of the MG.

Cornering is better than the skid-pad figures would indicate but becomes unpleasant on poor roads. The TR-6's handsome perforated disc wheels are fitted with very tall tires by current standards but the generous tread width makes a needed contribution to the car's handling. The low engine speed would make cruising very pleasant if wind noise (mostly from the top) were not so great.

The two words we'd choose for the Triumph are **Strong** but **Dated.** Despite the extremely intelligent modifications Leyland has made, the design needs to be replaced soon. The superb engine asks for a wider, unitized structure and improved suspension to go with it.

MGB Mk II

THE B-SERIES MG has had a long, successful life (since 1962) but—even more than for the Triumph—the end is in sight. Basically, the MGB Mk II (the new designation denoting minor styling changes) is a very good car. It has few serious faults but was consistently unimpressive compared

FOUR SPORTS CARS

to the outstanding qualities of one or more of the other cars in each category on our score sheets. The updating of the car—except for the much-needed all-synchro gearbox fitted two years ago—has been so stingy that it simply looks and feels old. Its price—lowest of the four cars—is therefore an important factor.

The MGB has enough power and torque to perform well in a straight line and it does much better than the Triumph on overall handling and ride though steering is notchy rather than progressive. Braking and fuel economy are below average for the group but still satisfactory. The new gearbox compared well with the others and earned the car a "best" vote from one scorer.

If the MG functions decently, it falls down on aesthetics and convenience. The styling has suffered rather than benefited from the recent modifications, while the interior is decidedly unattractive. In most cases, needed improvements have been made as cheaply as possible (for instance, the front of the hood still has the raised form for the MG badge which has been moved to the center of the grille, the rear bumper has been sliced apart to make room for a lower license plate mounting, the old dash structure has the new safety padding built on top of it, etc.). Carrying capacity is minimal, with the smallest trunk and only the tiniest pocket for incidental items in the cockpit. The fussy top design, requiring disassembly and separate storage, is Early-Masochistic.

Capable but **Undistinguished** would characterize the MGB. Its low price may keep it on the market for some time but it is badly in need of replacement.

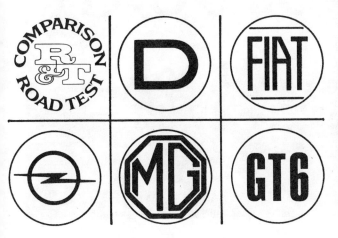

Comparing the Datsun 240Z, Fiat 124 Sports, Opel GT, MGB GT and Triumph GT6—a closer contest than we expected

WHAT DOES ONE get when he buys $3500 worth of Grand Touring car? Generally, a small, light closed car, adequately but not spectacularly powered, a cut above the average in handling and braking, offering a measure of comfort not found in an open sports car. This month we take a comparative look at five such cars. None of them is a new model, all having been tested before by R&T in separate road tests. There have been detail changes in all of them since we last tested them although their basic character has not changed; still there's nothing like getting a group of cars together, taking a journey in them and comparing them nose-to-nose. We're always surprised at how much we learn in a comparison test and the reader may be surprised at some of the results of this one. We were.

As the General Data table shows, the list prices of the five cars all cluster around $3500. And each car is reasonably complete at the basic price—you don't need to pay extra for adjustable seats, a cigarette lighter or a 4-speed gearbox. All are front-engine, rear-drive cars, all have inline piston engines, all have at least two disc brakes out of four, and all weigh between 2000 and 2500 lb; one is from Japan, one from Italy, one from Germany and two from England. And they are all very different from each other.

The Datsun 240Z set U.S. motoring on its ear when it appeared in early 1970. It seemed too good to be true: a really fast, good-handling and good-looking coupe with great refinement and extensive standard equipment—all at what seemed an incredibly low price. Strictly a 2-seater but a roomy one, it is the longest car of the group though not the widest, the heaviest by a small margin, and by leaps and bounds the most powerful with 150 bhp from its 2.4-liter overhead-cam 6-cyl engine. It has independent suspension all around (by struts and coils at both ends) and a combination of disc front brakes and drums rear. Its styling is professional and up-to-date if a bit "pop culture," and that the 240Z is an exciting package cannot be denied.

In fact, it is so exciting that it has generated a supply-and-demand situation virtually unprecedented in the U.S. The waiting list for delivery on one is as high as six months, even though it has been over a year since the model was put on sale. Datsun programmed a supply of 1600 per month for the U.S. and they're now getting over 2500, but they have found that the demand is for about 4000 per month. Interesting, for we asserted in the 240Z introduction story that a domestic carmaker probably could build an equivalent car for the same price ($3600) in quantities of 50,000 per year or so. Anyway, Datsun stopped production on the 1600 roadster to make way for more 240Zs but the supply is still far behind. In some areas, for instance, dealers are telling customers they can't get the cars without lots of optional equipment—wide alloy wheels, air conditioning, etc., and getting away with it. And the *Kelley Blue Book* retail value for a used 1970 240Z is over $4000!

So—though the list price of a 240Z may be only $3596—one may or may not be able to get one for that price. If you can find a dealer who will sell you one for that price you'll probably have to wait months to get it. We must give Datsun the credit for producing such a package at such a reasonable price but we have to caution the reader that Datsun's price may not be the dealer's price. The laws of supply and demand still work, list prices or no.

Standard equipment on the 240Z includes a signal-seeking AM radio (with electrically powered antenna) and a heated rear window. Our test car had no options except a set of wide wheels (14 x 5½, part no. 40300E4600, $13.50 each) that can be obtained from Datsun dealers but are not installed at the factory; for the price table we estimate a total charge of $74 for the wheels and installation.

The Fiat 124 Sport Coupe has been updated this year with a smooth new front-end look and a 1608-cc engine of longer stroke than the older 1438-cc unit. It's not any faster than before but the larger engine is stronger in the middle ranges (so that less gearshifting is required to maintain a given pace) and both smoother and quieter. The Fiat is almost as long as the Datsun and somewhat wider; it is the only car in this group to offer a real rear seat and it's so real a seat that the car can be compared to some small sedans. It has several engineering distinctions: the only twincam engine in the group (the two camshafts are driven by a single toothed belt), the only 5-speed gearbox and the only 4-wheel disc brakes. It's the only car in the group with a separate, lockable trunk where one can hide valuables; three of the others have "tailgates" and one a cockpit luggage area. And its luggage capacity is the largest in the group.

The 124 Coupe test car, working from a POE price of $3292 (the lowest in the group), had the nice Cromodora alloy wheels, which cost only $135 for the set, a rather poor AM/FM radio (one shouldn't expect much at $85) and add-on chrome side strips and luggage rack; the last two items aren't included in our price tabulation.

British Leyland's MGB GT has been around since 1966

THE $3500 GT

and is based on the MGB roadster introduced in 1962. It's a classic British sports car but with a nicely designed fixed roof. The only one of the group attempting to be a 2+2, it has a very small bench seat behind the two main ones; this is big enough for small children only, and it can be folded to extend the luggage area. Surprisingly, the B is nearly as heavy as the Datsun, and with only 92 bhp from its pushrod four it is the slowest of the lot. Its front suspension is conventional—unequal-length A-arms and coil springs—and rear suspension is by the simplest means possible, a live axle on leaf springs.

The MGB GT comes with radial tires as standard equipment; styled steel wheels are standard, wire wheels (either painted or chrome) optional. Our car had the optional overdrive ($165), the familiar Laycock-de Normanville hydraulic unit that shifts in or out at a flick of a stalk on the steering column to give a 0.802:1 reduction. Thus 4th gear becomes 3.14:1 overall instead of 3.91:1 and the MG becomes the longest-legged of the lot. The overdrive works on 3rd gear also.

Opel's GT is based on the Kadett chassis, so it actually isn't as up-to-date in that department as its bulkier cousin the 1900 Rallye. Its front suspension is odd, based on a

transverse leaf spring and one set of lateral control arms, but at the rear everything is shipshape with a live axle located by trailing arms and Panhard rod and sprung by coils. Its all-steel unit body (many people assume it's plastic because it looks so much like a Corvette) is made in France to a very high standard of finish and for 1971 it is available only with the 1.9-liter engine rather than both the 1.1 and 1.9. The engine has taken a power cut, though, to satisfy the politicians; bringing the compression ratio down from 9.0 to 7.6 so it could run on 91-octane fuel has reduced power from 102 to 90 bhp. Performance has suffered (0-60 time is up from 10.8 to 11.9 sec) but the GT still holds its own in the class. Amazingly, the GT has had a price cut too—it's now nearly $200 cheaper than it was in 1969 with the 1.9 engine.

The Opel GT is strictly a 2-seater: luggage is carried on a flat floor behind the passengers and loaded in through the car doors. The spare tire lives behind a vinyl partition just aft of the luggage area. Thus it has the least convenient luggage accommodation, though not the least capacious. Other distinctions for the Opel: it's the lightest and most economical of fuel in the group and its brakes can stop it in the shortest distance. Our test car had but one extra, a good AM

PHOTOS BY GORDON CHITTENDEN

Datsun's engine is the largest and its instruments and controls the best.

Fiat has the only dohc engine and excellent instrumentation.

Opel's engine has generous displacement and unusual high-cam design.

MG has time-proven pushrod engine, safety-modified instrument panel.

GT6's smooth engine is very accessible, its wood-panelled dash luxurious.

$3500 GT

radio at approximately $75 (dealer installed); like the Datsun, it can be ordered with a 3-speed automatic transmission.

Before the 240Z appeared the Triumph GT6 was the only car in its class with a 6-cyl engine, but now its 95-bhp, 2-liter six is no longer a big attraction. It's the smallest car of the group and only insignificantly heavier than the Opel; these points relate to the fact that it's derived from Triumph's smallest, lightest sports car, the Spitfire. It shares basic chassis structure (a backbone frame with separate body), front suspension and body structure from the beltline down with the open-bodied, 4-cyl Spitfire though the Spitfire's swing-axle rear suspension is replaced in the GT6 by a more satisfactory unequal-arm linkage for the rear wheels. The GT6 has the tightest interior dimensions of the group but not the smallest luggage capacity, and its luggage area is loaded easily through a tailgate as in the Datsun and MG.

This year the GT6 has been freshly restyled on the outside, and if some staff members noted that the rear end is reminiscent of the old Sunbeam Harrington coupe, all agreed that the car is better looking than before. It has one feature that makes it uniquely maneuverable in a crowded city: a tiny turning circle made possible by front wheels that can be steered well beyond the limits of proper geometry. One can turn around in just 25 ft, 6.5 ft tighter than the next twistiest car in the group, the Datsun.

Standard equipment at the GT6's basic price of $3424 includes white-stripe radial tires and a heated, tinted rear window; it has a 4-speed gearbox and an overdrive like the MG's can be ordered as an option; brakes are a disc/drum combination. Our test car had only an AM/FM radio—again a rather poor and low-priced ($100) one. A final distinctive feature of the GT6: A rich wood-panel dashboard in the traditional British manner.

As IN PAST comparison tests, we chose a test route appropriate to the character of the cars. In this case it was a route we'd used two years ago for a group of more expensive GT cars. We left our office in Newport Beach, topped up all the cars at a nearby filling station, and drove south

GENERAL DATA: 5 MEDIUM GTS

	Datsun 240Z	Fiat 124 Sport Coupe	MG B GT	Opel GT	Triumph GT6 Mk3
Basic POE price*	$3596	$3292	$3620	$3306	$3424
Price as tested	$3745	$3562	$3823	$3409	$3674
Engine position/driven wheels	f/r	f/r	f/r	f/r	f/r
Chassis type	unit	unit	unit	unit	separate
Brake type	disc/drum	disc	disc/drum	disc/drum	disc/drum
Swept area, sq in./ton	233	227	227	222	209
Suspension, front	ind. coil	ind. coil	ind. coil	ind. leaf	ind. coil
rear	ind. coil	live coil	live leaf	live coil	ind. leaf
Standard tires	175-14 rad	165-13 rad	165-14 rad	165-13 bias	155-13 rad
Steering turns, lock-to-lock	3.5	2.7	2.9	3.0	4.5
Steering index	1.10	0.99	0.93	0.99	1.13
Fuel tank capacity, gal	15.9	11.8	12.0	13.2	11.7

*POE prices vary slightly for east, west and Gulf ports
As-tested prices include: for Datsun, 14 x 5½-in. wheels and installation; for Fiat, alloy wheels, AM/FM radio; for MG, overdrive; for Opel, AM radio; for Triumph, AM/FM radio. All as-tested prices include charge for preparation at dealer.

Opel's separate lap-shoulder belts work but are a mess.

on the Coast Highway through Corona del Mar and Laguna Beach (stoplights galore) to Dana Point, where we turned inland to connect with Ortega Highway, California 74. This highway, a nicely surfaced, 2-lane route with a delightful variety of turns and hills as it winds through some of Southern California's finest country, was lightly traveled and the weather was beautiful; it was easily the highlight of our trip. At Lake Elsinore we connected with route 71 toward Temecula; after Temecula, route S16 to Pala, 76 to Santa Ysabel and 78 to the little town of Julian high in the Laguna Mountains, where we stopped for lunch. Then on down 78, another wonderfully twisty road, into the Anza-Borrego desert, out across the desert at speeds dictated by road conditions and car capability rather than artificial limits, up through the Joshua Tree National Monument, and over clear, generally straight back roads to our overnight stop at Victorville, from where we freewayed it back to Newport Beach the next morning. In all, a rich and varied 500 miles of motoring in which we found out all about the five GTs.

Back at the office we set about scoring the cars. Each driver was given a score sheet on which he could rate each car on 15 different aspects of behavior, such as handling, ride, quietness, braking, steering, gearbox, engine, controls, seating, ventilation and heating, vision, finish, luggage accommodation and so forth. All these categories could be scored on a 1 to 10 basis, 10 being the score for a topnotch performance and 1 being the lowest score possible. These scorings were then totaled for each driver and for the entire group to get an overall rating score for each car.

In addition, each driver was asked to rank the cars in the order of his personal preference—disregarding, if need be, his separate ratings of the car's various aspects. Here's how the ratings turned out:

The Datsun scored the highest point total. In individual driver scoring, it garnered the highest number of points from four of the five drivers, and three of the five drivers rated it their personal favorite.

Next came the Fiat, and here we emerged somewhat surprised. It had been generally anticipated that the Datsun would win by a large margin, but not so. The Fiat tallied an impressive score, little less than the Z; one driver gave it more points than he gave the Datsun, and the same driver gave it his personal nod.

Then the Opel. It was a clear step below the Datsun and Fiat but clearly not in the doldrums. One driver rated it his personal favorite, though in scoring he had given the Datsun more points.

In total points the MG was not as far below the Opel as the Opel was below the leaders, and there was no unanimity in the personal ratings of the MG by the various drivers: two rated it third, two fourth and one last. But these ratings averaged a 4th-place finish just as clearly as the points score indicated; in fact, averaging the "position" of each car over the five drivers' listed orders of preference, the cars stacked up the same way: Datsun, Fiat, Opel, MG, Triumph. Which brings us to the Triumph: It came in last, but not far behind the MG and was ranked last by four of the drivers on their personal ratings. The one driver that ranked it 4th instead of 5th also gave it more performance points, so it had a clear attraction for him. Now let's look

MG's tailgate-loading luggage area is handy.

GENERAL SPECIFICATIONS: 5 MEDIUM GTS

	Datsun 240Z	Fiat 124 Sport Coupe	MG B GT	Opel GT	Triumph GT6 Mk3
Curb weight, lb	2355	2220	2345	2110	2115
Test weight, lb	2770	2620	2725	2500	2490
Distribution, f/r, %	51/49	55/45	49/51	54/46	54/46
Wheelbase, in	90.7	95.3	91.0	95.7	83.0
Track, f/r	53.3/53.0	53.0/51.8	49.3/49.3	49.4/50.6	49.0/49.0
Length	162.8	162.3	152.7	161.9	149.0
Width	64.1	65.8	59.9	62.2	58.5
Height	50.6	52.8	49.4	48.2	47.0
Luggage capacity, cu ft	8.5	9.6	6.3	6.6	6.6

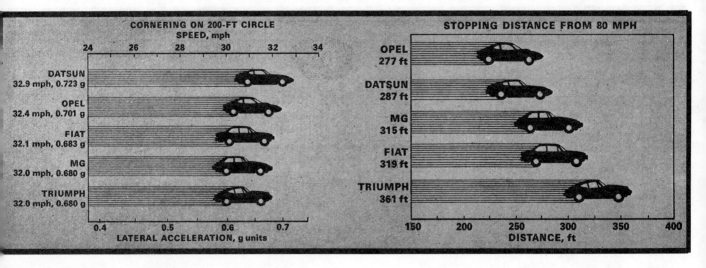

CORNERING ON 200-FT CIRCLE
SPEED, mph

DATSUN 32.9 mph, 0.723 g
OPEL 32.4 mph, 0.701 g
FIAT 32.1 mph, 0.683 g
MG 32.0 mph, 0.680 g
TRIUMPH 32.0 mph, 0.680 g

LATERAL ACCELERATION, g units

STOPPING DISTANCE FROM 80 MPH

OPEL 277 ft
DATSUN 287 ft
MG 315 ft
FIAT 319 ft
TRIUMPH 361 ft

DISTANCE, ft

THE $3500 GT

at the cars in detail and in the order of their ranking.

Datsun 240Z

IN GENERAL the 240Z lives up to its promising specification. Its generous-size engine delivers the smoothest, quietest and dramatically the most powerful performance in the group; it is an engine that pulls strongly from low speeds, runs silently at high cruising speeds and continues to be impressive right up to its 6500-rpm yellow line on the tachometer, thanks to improvements in the crankshaft since our earlier road test. All this performance isn't without cost; the "Z" is also the thirstiest in the group, but 21 mpg is nothing to complain about and the fuel tank is large enough to give it a cruising range to match its cruising ability.

But the Z-car has a serious problem associated with high-speed motoring. It's very sensitive to sidewinds at speed, and when traversing road undulations at high speed (as we did repeatedly on the desert highways) it requires a lot of motion at the steering wheel to keep it on course. Datsun has some chassis tuning work to do here, and in the meantime an owner can fit an undernose spoiler, available from BRE (137 Oregon St., El Segundo, Calif. 90245) for $32.

In low- to medium-speed cornering and handling, however, the 240Z shines. Surprisingly, the 5½-in. rims didn't make a difference in absolute cornering power over the standard 4½-in. ones, though they added crispness to the car's response; but it still led the group with a 0.723g cornering capability. And though its steering is not the most pleasant or accurate in the group, it is quite acceptable, never unduly heavy and certainly quick enough.

The Z's combination of front discs and rear drums, vacuum assisted, tie with the Opel's brakes for best-in-group; though the "panic" stopping distance from 80 mph takes 10 ft more than the Opel, the car stays under control a little better under these conditions and fade under hard repeated use is negligible.

Comfort and accommodation also rank high in the 240Z. It accommodates only two people, but what space for those two! The largest driver in our test crew, who is 6 ft 2 in. tall and weighs 200 lb, gave it the full 10 points on every aspect of its interior but finish—undeniably some of the materials used, notably the quilted-pattern vinyl, are less than pleasing. Controls are notably good, everything being within good reach for a typical male driver with his 3-point belt fastened (the best belt in the group, by the way), and the steering column-mounted lighting control rates special mention. On the other hand, it was the only car in the group

without some provision for easy daylight headlight flashing, and the ventilation system, though it provides a good flow of outside air that can be boosted by the blower, unfortunately aims most of it right between the driver and passenger.

In sum, the Datsun 240Z's plusses are its striking looks, its effortless, strong performance, its good brakes and low-speed handling, and its comfort and equipment. On the negative side, the only serious criticism is about the high-speed stability. If you can get one for list price, or even get one with the extras *you* want, it is not only the best car in the group but the best buy.

Fiat 124 Sports Coupe

THE FIAT deserves more popularity. At nearly $200 less than the Datsun with comparable equipment, it did so well in our comparison test that it scored nearly as many points. Of course it doesn't offer the zoomy styling of the Datsun (the boxy shape that turns it into a true 4-seater doesn't allow that) nor the brilliant performance. Its 4-cyl engine, the smallest of the group, is nevertheless a most satisfying bit of machinery: quiet, very smooth for a 4-cyl (easily the best four in the group), and willing to rev happily to its 6500-rpm redline. And the 5-speed gearbox is the best gearbox in the group.

In road behavior the Fiat scores at the head of the group. Its steering is the most precise, its handling the best; it really shines at high speed in contrast to the Datsun, for it isn't blown about by sidewinds and can negotiate high-speed dips and humps without a hint of losing its composure. Only an over-eager vacuum brake booster detracts from its overall roadworthiness; our drivers always found themselves overdoing it with the brakes when first getting into the car. And though it's the only car in the group with disc brakes all around, it doesn't do anything impressive in a panic stop and the brakes squeal often when not in use. Its brakes do have the best fade resistance in the group, though.

The Fiat's driving position is, in a word, odd—and perhaps something we, as Americans, will never understand. The steering wheel is buslike in that it is less vertical than usual, and it is far away from the driver while the foot pedals are close. But the seats are good and so are the controls, which make maximum use of modern steering-column stalks to do various things. The seatbelts were installed incorrectly on the test car so that either lap or shoulder section was twisted. Ventilation is not particularly good, but ventwings in the doors make it possible to drive at moderate speeds with the door windows open and no drafts. The Fiat has the best vision outward of any car in the group, so it's a good car for city traffic.

With its good rear seat and capacious, separate trunk the 124 Coupe is far and away the most practical car of the

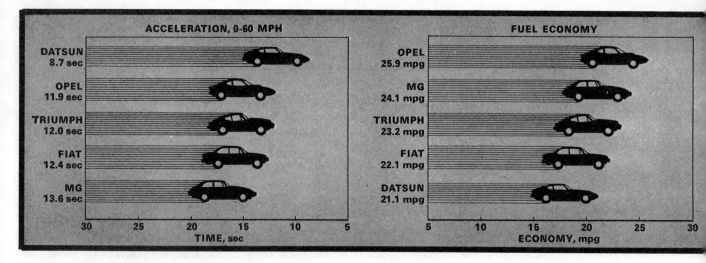

ACCELERATION, 0-60 MPH

	Time
DATSUN	8.7 sec
OPEL	11.9 sec
TRIUMPH	12.0 sec
FIAT	12.4 sec
MG	13.6 sec

TIME, sec

FUEL ECONOMY

	Economy
OPEL	25.9 mpg
MG	24.1 mpg
TRIUMPH	23.2 mpg
FIAT	22.1 mpg
DATSUN	21.1 mpg

ECONOMY, mpg

group if one needs more than 2-passenger accommodation. And surprisingly it doesn't give up a thing in sportiness for this extra measure of utility; in fact, it is *the* driver's car of the group. It is also well finished and equipped. On the minus side are its modest performance, grabby brakes and fuel economy that isn't impressive. We placed the Fiat 124 Sports Coupe as a close second in the group.

Opel GT

THIS ONE, coming in third in the group, is a crisp little package but not an impressive value for the money. Regardless of what one may say about its pseudo-Corvette styling, it is extremely well finished and put together and ranks fairly high in its driving position and control layout. It has the 2nd-best gearbox, light if not dead-accurate steer-

ing and the highest performance level of "the others" (i.e., other than the Datsun) despite its excellent fuel economy. It's the only one of the group that uses regular fuel.

The body is structurally solid and rattlefree. Vents at the dashboard ends provide a good flow of ventilating air, and longlegged gearing makes the engine (which is rather hashy sounding up through the gears) fairly quiet at speed in 4th; thus one can cruise with the windows up even in warm weather at an untiring noise level. On curvy roads the GT is amazingly well-planted and stable considering its humble origin, but in hard low-speed turns the rear axle chatters and its general handling characteristics are too much on the understeer side for maximum entertainment value. As we said in our road test of it, it may not be a really good-handling car but it is safe and predictable.

Fiat offers the only 5-speed gearbox in the group.

ENGINE & DRIVE TRAIN: 5 MEDIUM GTS

	Datsun 240Z	Fiat 124 Sport Coupe	MG B GT	Opel GT	Triumph GT6 Mk3
Engine type	L6 sohc	L4 dohc	L4 ohv	L4 sohc	L6 ohv
Bore x stroke, mm	83.0 x 73.3	80.0 x 80.0	80.3 x 89.0	93.0 x 69.8	74.7 x 76.0
Displacement, cc	2393	1608	1798	1897	1998
Bhp @ rpm	150 @ 6000	104 @ 6000	92 @ 5400	90 @ 5200	95 @ 4700
Torque @ rpm, lb-ft	148 @ 4400	94 @ 4200	110 @ 3000	111 @ 3400	117 @ 3400
Transmission	4-sp man[1]	5-sp man	4-sp man[2]	4-sp man[1]	4-sp man[2]
Standard final drive ratio	3.36:1	4.10	3.91	3.44	3.27
Engine speed @ 70 mph, rpm	3350	4000	3980	3560	3480

[1] 3-sp automatic optional   [2] overdrive optional

Datsun's performance is in a class by itself.

PERFORMANCE: 5 MEDIUM GTS

	Datsun 240Z	Fiat 124 Sport Coupe	MG B GT	Opel GT	Triumph GT6 Mk3
Lb/hp (test weight)	18.5	25.2	29.6	27.8	26.2
Top speed, mph	122	112	105	110	107
Standing ¼ mi, sec	17.1	18.6	19.6	18.4	18.6
Speed at end of ¼ mi, mph	84.5	72.5	72.0	74.0	74.5
0–60 mph, sec	8.7	12.4	13.6	11.9	12.0
Brake fade, % increase in pedal effort in six ½-g stops	10	nil	17	nil	20
Stopping distance from 80 mph, ft	287	319	315	277	361
Control in 80-mph panic stop	excellent	very good	fair	good	good
Overall brake rating	very good	good	good	very good	fair
Actual speed at in. 60 mph	61.5	56.0	58.0	56.0	58.0
Fuel economy, mpg (trip)	21.1	22.1	24.1	25.9	23.2
Cruising range on full tank, mi	335	260	290	340	270

THE $3500 GT

In recognition of the value of seatbelts—and in the hope that we can influence more people to use them—we're rating the belts in each car. Opel has followed U.S. GM practice by simply fitting separate lap and shoulder belts, each with its own pushbutton buckle. These fit well once in place and the roof anchorage for the shoulder belt is far superior, for instance, to the Fiat's which is on the body side behind and slightly below (that is bad—collarbones can be injured) the shoulder. But the separate belt arrangement makes it extremely inconvenient to strap oneself in, thus making it less likely a driver or passenger is going to do it.

Another safety-related item: throne-type seats used to meet the federal government's head-restraint rule impair vision to the rear, and a blind spot created by the rear roof is a further vision problem. The GT needs more outward vision for traffic maneuvering.

But all in all, it's a pleasant if not exciting little coupe. Not a bad car at all—it's just that the Datsun and Fiat are so good.

MGB GT

WE'VE HEARD that British Leyland is simply letting the MGB run its historic course; when it can't be sold anymore they'll drop it and that's that. The car seems to bear it out. Meeting the U.S. crash-safety regulation was done by laying an ugly, add-on instrument panel over the existing one and the little bit of styling facelift has been done in a haphazard way.

It's truly a car of the past. Everywhere there's evidence of a sports car designed and built in the traditional manner—in a rather homemade way, to be blunt, in great contrast to the professional design and execution of the Datsun, Fiat and Opel.

That impression carries through on the road. The GT is heavy (nearly as heavy as the Datsun) but gets only 92 bhp from its noisy pushrod engine; so it's the slowest of the group. And a rather balky shift linkage doesn't contribute to driving fun—a surprise, because this was one area in which MGs always excelled in the past.

The optional overdrive does make the B GT a capable long-distance tourer; in OD at 70 mph it's turning 3190 rpm, vs the 3980 given in the Engine & Drive Train table for normal 4th gear. And the overdrive gives it the second-best fuel economy figure. But don't expect the MG to be quiet at speed even with overdrive; there's so much wind noise you'd think it was a roadster, not a coupe.

The B handles well enough but rides very stiffly. At the limit there is a bit of oversteer that makes it fun to toss the car around, especially on low-speed curves. The steering is heavy, but the MG has the quickest steering in the group. The brakes are about average.

Ventilation, provided by a simple flap under the dash, is ineffective compared with the best of the group, but one can at least maximize it by opening the door ventwings and/or the swing-out quarter windows.

Vision outward is quite good, and MG augments it with a curious righthand fender mirror, stuck out there all by itself. The seatbelts are of the simple Kangol variety, same as on the Fiat, and someone at the factory or distributor had also installed them wrong so that one section of the belt had to be twisted.

There's little to redeem the MGB GT, not even a low price, and we can only call it a holdover from another era.

Triumph GT6 Mk 3

THE GT6 IS almost as close to the MG in our score-giving as the Fiat to the Datsun. It rates close to the Opel in performance, and with six cylinders its engine is smoother and quieter than all but the Datsun and Fiat. If overdrive is ordered, Triumph installs a 3.89:1 final drive rather than the 3.27:1 of the test car and in this form it will be a bit quicker through the gears. But what promises to be a good open-road car (if the smooth, adequately powerful engine is any indication) turns out not to be because a drumming driveline vibration sets in at about 65 mph and stays there as speed rises.

The gearbox is stiff-shifting and the shifter's H-pattern is oddly skewed; this all takes some getting used to and perhaps the owner could adapt. In any case, the GT6 is a decently enjoyable car over a curvy road at moderate speeds, with light if not particularly quick steering and good handling response. But in ultimate cornering power it is at the bottom of the group. Don't expect the brakes to accomplish much—though their fade resistance is adequate they take a very long distance to stop the car from 80 mph. On bumpy or irregular road surfaces the GT6's backbone-plus-body structure is the least staunch in the group, creaking and rattling when the going gets rough.

There are charms to the GT6. Its interior materials look the richest of any in the group and the instrument layout is particularly handsome. The seats have been upgraded in recent years (as have the MGB's), but in the GT6 they have "throne" backs like the Opel's which are a bit restrictive for rear vision. Also like the Opel, the GT6 has a blind quarter that hampers one's ability to maneuver freely in traffic.

Though the GT6 scores esthetically and in performance over the MGB, it loses it all on comfort. Its seating is the most cramped in the group, the steering wheel is very high, and its seatbelts were next to impossible to adjust to fit anybody. Triumph has gone to greater lengths to update the GT6 than MG has the B, but it still failed to make much of an impression on R&T's five testers and had to be rated last-in-group.

ONE OF OUR five drivers commented after the trip that the Datsun should be rated separately, as it is simply a class above the rest. But when all was said and tallied, the Fiat came surprisingly close and the Opel was far from unpleasant. As for the two Britishers, we do not wish to kick dead horses and sincerely hope that England will be able to get off her duff, produce some competitive cars again and challenge the other countries. We have reason to believe that British Leyland does intend to keep building sports cars and to come up not only with new designs but to realign the product "mix" of MG, Triumph and Jaguar. One of these new products, we would predict, will be a medium-price GT replacing both the B and the GT6—one that we hope will render the choice of a good $3500 GT a bit more difficult to make.

ALLAN GIRDLER PHOTO

M Y PARTICIPATION IN the New England "T" Register's Marathon was assured when TC owner Mike Williams telephoned for advice: Do I know anything about TC differentials?

"Swill me in a shallow pan of petrol if I don't," I said, quoting from the owner's manual for emphasis. "Working on TC differentials used to be all I ever did."

"My car goes 'click, click' when I shift."

"They all do. Nothing to worry about. Uh, it doesn't go 'clank, clank' does it?"

JOUNCING TOWARD ABINGDON

BY ALLAN GIRDLER

1000 non-stop miles in a TC, as seen (and felt) by a man who thought he knew better

"A friend of mine has a spare differential he'll lend me for the trip."

"Bring it along."

I thought I was through with all that. In 1955, I bought an MG-TC. During the 14 years that followed I raced it, rallied it, drove close to 150,000 miles on the highway, literally from coast to coast. I got frostbitten in New York and watched passengers pass out from the heat in Texas and Kansas. I replaced every part in the car at least once and learned that I am so-so at upholstery and terrible at carpentry. Enjoyable, all of it, but when the car drove away without me in 1969 I cannot say my heart was pierced by more than a tiny twinge. And I'm not sure there was even that.

Nostalgia wasn't a factor, then, in my enlistment for the marathon. What appealed most was a certain specious logic.

Primary sponsor is the New England "T" Register, an organization based on T-series MGs. Not a club, they say, because the members aren't people. One registers one's car. The car is the member. The owners (the owned would be just as accurate a description) tag along. The register has enjoyed a bewildering success. The co-founders expected maybe 150 owners to sign up for a newsletter, parts exchange, etc. At last count, there were 2000 MGs of TC, TD and TF designation enrolled.

Twice each year the Register hosts a get-together. The proper name is Gathering of the Faithful. The site usually is a resort hotel in New England. T-series MGs come from all over for rallies, gymkhanas, concours, parts swaps and a weekend of mutual admiration.

Parallel to the Register is the Vintage MG Car Club of Chicago. Same goals, same type of MG, a good share of dual registration.

Now. It occurred to Herb Nichols, TD owner and mem-

ABINGDON

ber of both clubs, that several other dual members drive to the Gatherings. T-series people and cars are competition minded. And there aren't many places for them to compete. Why not, he reasoned, combine all this into a marathon, a timed drive from the midwest to the GOF, a 1000-mile test, roughly (interpret that any way you wish), of man and machine. Nichols drew up some rules, secured Register support and enlisted entries. The first marathon was run in 1968, with each successive run attracting more and better-prepared cars.

Eighteen cold, wet hours later, the windshield was still down.

The Marathon isn't a rally; the routing is flexible, the winners picked on elapsed time. Nor is it a race; most of the roads are major highways, with limits above the practical cruising speeds of the competitors.

The Marathon is an Act of Faith, a pilgrimage to a spiritual Abingdon, to a cool little corner of New England where the year moves only from 1948 to 1954 and back again.

The object is to drive from a pre-determined point to the GOF and to do it in the least possible time, stopping only for fuel and mechanical derangements. (All former T-series owners pause here and reflect. An Act of Faith, indeed.)

There are two major rules: T-series MGs only and no non-T engine swaps. Winners are determined by elapsed time multiplied by a handicap. Scratch car is the TF 1500, in theory the T-series best suited to fast—well, relatively fast—cruising. The older the car and smaller the engine, the larger the handicap. The intent is to encourage the older cars to compete and to not lose heart when newer models whiz past.

The first marathons started from the midwest. For 1971, by popular demand, there were starting points east and west, so New England-based faithful could participate.

The man for whom I crewed didn't actually need much help. Mike Williams bought his TC in baskets and restored it, after practicing on two Austin-Healeys and an MG TF. Nor did he bring the differential. After stowing the sidecurtains, tools, gaskets, a spare half shaft, oil, gear lube, first aid kit and suitcase, there wasn't any room. Our provision for the trip was a sack of turkey sandwiches: "I asked my wife what will keep for a long time without refrigeration,

and she said turkey would keep the longest." So turkey it was.

To date, no one has actually built an MG for the marathon, but the regulars do equip their cars with the event very much in mind. Nichols' car looks like the modified TDs of 20 years ago, with 8-port aluminum head, hood strap replacing the side panels, velocity stacks in place of air cleaners, etc. Another TD was supercharged. A third has an MGA rear axle, complete with 4.3:1 ring and pinion. Doesn't climb hills very well, he said, but it surely helps on the flat. The only pure-stock TD in the western contingent did what he could and ran the distance with windshield folded to minimize wind resistance.

But the rival we feared most created terror by his very lack of preparation. Owner Bob Pickard drove his TC, Charley, 700 miles from home to starting line. Charley has 247,000 miles on the odometer, sounds like a sack of nails and holds the TC record for the marathon. A clear case of supernatural forces at work.

The western starting point was a gas station at the western edge of Michigan, adjacent to an Interstate that leads to Detroit. The route was across Michigan to Windsor, Canada, and on to Toronto and Montreal. From there we could go east in Canada and veer due south to the gathering at Waterville Valley, New Hampshire. Or an entrant could head south from Montreal and work east through Vermont.

Our TC had the lowest serial number, so we were the first car away. The start was at night, in order, I guess, to have the teams fresh during the darkness while being deluded out of fatigue by the rising sun. Which is how it works out, I found.

Mike took the first stint behind the wheel. Mind, I didn't just sit there. The copilot's duties included handing over turkey sandwiches, shining a flashlight on the coolant gauge mounted in the radiator cap and on the oil pressure gauge, located in the dashboard but minus illumination for some obscure electrical reason, and—most important— looking back to see if anything was gaining on us.

We had the idea, correct as it happened, that we were the slowest car in the western group. The entrants left at 5-minute intervals. If we were passed by the second car in one hour, we reckoned, that would mean he had 5 mph on us, and so on down the list.

An uneventful hour passed. Then another, while Mike sawed at the wheel and I peered at the gauges. Three hours out, the marathon turned into a dead heat. Charley went by, with the first TD a minute behind him, the second TD a minute after that and the supercharged TD two minutes after that. They all waved.

Very discouraging. Without consulting me, Mike and Pickard had agreed to dispense with the handicap. The TCs would face the TDs even up. We were the slowest, and all we had to look forward to was mechanical troubles for

Owner/driver Mike Williams and his triple-threat TC.

JIM WINDMEIER PHOTO

ALLAN GIRDLER PHOTO

Robert Herlin's supercharged K-3 Magnette was the star of the gathering.

everybody else, a prospect not conducive to sportsmanship.

For example: All the cars were close together at the first fuel stop, save one TD which evidently had taken a wrong turn. We all laughed.

My turn. Driving a TC again felt familiar, in some ways. The tiny pedals, the upright posture, the flexed elbows were just as I remembered them. By the time I sold my TC, though, it had undergone modifications that provided half again as big an engine and exactly twice the stock power. I had forgotten how much of the TC's wonderful exhaust was brag, and how little of that powerful roar was fact.

And the steering! Terrible! In top condition, TC steering is bad enough. With all possible modifications, it's barely tolerable. This car had a bad unit. Mike adjusted, to no avail. Either it was heavy, vague and notchy or it was heavy, vague, notchy and sloppy.

We kept Charley in sight across Detroit and through the tunnel to Canada, but then he pulled away. Past midnight, now, and getting colder. We unfolded the windshield at dusk, put up the top when it was truly dark. Mike burrowed through the parts and equipment, hauled the front side-curtains out of their bin and installed them, all on the move. Great skill involved there, although maybe you'd have to try it to appreciate it.

All my old skills came back. Stuffing the left hand into the heater, winding the right hand through the spokes of the steering wheel and stuffing it into the other heater outlet. Jamming the left leg between clutch pedal and transmission tunnel so as to unbend the knee. Folding the left leg between gearshift and seat cushion when heat from the transmission blisters the foot.

I was reminded of Le Mans—no, wait! Guiding a TC at 65 mph must require the same degree of skill and concentration as guiding a 917 at 240 mph. True, the MG gives you more time to keep between the curbs but you need a much larger arc of correction.

And at Le Mans when one's driving stint is done, one climbs out of the car, withdraws to a quiet room and lies down on a soft cot. In the Marathon the co-driver is still plugged in, bolt upright, wind chilling his kidneys, exhaust howling in his ear, the water temperature and oil pressure gauges still demanding his attention, the driver still demanding turkey sandwiches. (Actually, I just said that for effect. The demand for turkey sandwiches slackened considerably after the first dozen.)

During my tenure of ownership I did just about everything that can be done in a TC. Sleeping wasn't one of them. Mike managed, somehow, to doze while curled up with his jacket over his head. "Ya slept all night," I said later. "I had my eyes closed for 40 minutes," he countered.

Do not get an impression of gloom and discomfort. The cheeringest things happened, like passing Charley, reduced to 55 mph by (we learned later) an oil-soaked clutch. As we passed them, I waved.

Returning to acts of faith for a moment, the supercharged TD had the only service car in the marathon. Somewhere west of Toronto a fan blade broke off and sliced through the lower radiator pipe. When the back-up car arrived it towed the TD to the nearest town, where the car was left for repairs. This was the only retirement of the event, that is, the man who prepared to have the car let him down *did* have the car let him down. The moral is obvious.

We caught the blown TD before his support car did. The crew was standing next to it, with disconsolation writ large

Marathon founder Herb Nichols, right, vs a frozen thermostat.

on their faces, allowing me to use that famous racing quip: "Nothing trivial, I hope."

Mike laughed and laughed. There's a fatigue factor here. Canadian winters buckle the pavement something fierce, and a TC goes over the buckles in great crashing leaps and bounds. So do the occupants. My image of us sailing topward in unison reminded me of my favorite Laurel and Hardy movie. I said as much. When we stopped to adjust the water pipe to the heater, Mike climbed out with, "Be right back, Ollie," and I laughed. It seemed funny at the time.

As the sky lightened, our spirits rose. At the 10-hour

ABINGDON

mark, we had covered exactly 570 miles. The TC course record was a 53-mph average, so we looked like contenders again. Especially when we passed the stock TD, hood open, occupants puzzled, just outside Toronto.

Another high point: A Canadian T-series fan provides coffee and doughnuts for the Marathoners. This year he was invited to crew a trans-Canada rally, which he did, leaving his wife to brew gallons of coffee, buy sacks of doughnuts and load them and three small children into the family station wagon at dawn and spend several hours ministering to passing Marathoners. It was indeed a welcome wagon. I am only sorry that several entrants weren't willing to use five minutes taking advantage of her hospitality. But the kids said they enjoyed eating one dozen doughnuts each.

Somewhere between Toronto and Montreal—I know, that's a long somewhere, but I was driving, not navigating—we caught one of the eastern contingent's TDs. They started half an hour before we did, so we were ahead, and craftily tucked in behind, as they had driven through Montreal previously. A good thing, too, because our map reading would have got us lost.

Outside Montreal we passed the last mandatory gate, a tollbridge. Here I made our Maximum Blunder. We had the choice of going east in Canada then veering due south to the center of New Hampshire, or going south from Montreal into Vermont and then east.

Mike had worked it out on the map, and wanted to go east then south. The TD headed south.

"Better roads the other way," Mike said.

"Those people have proved they know better than our maps," I said. My will prevailed, mostly because I was driving at the time, and I quickly wished it hadn't.

The land between Montreal and the ocean is covered with mountains and valleys, the latter running north-south. We got south quickly but then learned we had to meander over the hills, on narrow country roads.

Any other occasion, fine, but time was running out. Strange, the human mind. While we crawled across the endless plains, all we could think of was the machinery. For the last 100 miles the gauges, the pressures and temperatures, the odd noises were all forgotten. We had more important things to worry about. The 57-mph average fell victim to 40-mph corners.

We couldn't find the instructions in the jumbled equipment. We took a wrong turn. "Haw, haw," said a local with that rural good humor that makes big cities so attractive, "all you MG guys are a-gittin' lost."

Indeed. We couldn't even find the finish line. 'Round and 'round the hotel parking lot we whirled in a shower of mud and curses, furiously beeping the horn in the 3-long pattern supposed to notify the timers that we had, at last, arrived.

After one or two circuits, we followed the waving hands

to a table containing the scorers, who collected our toll tickets and clocked us in. The last 7 miles took 17 minutes. The finish line was hubcap deep in T-series cars. N realizing that only a fraction of the Register cars run i the marathon I thought we were among the final finishers. Not content to compare our efforts with Le Mans, I glumly thought of Indy, another place where Rookies Never Win.

There is a happy ending but first, the Gathering. What the Faithful mostly do when gathered is talk about and look at their own and each others cars. There was a rally, a concours d'elegance, a swap meet, a gymkhana and a raffle. All low in pressure. We missed the rally but I inferred that it was chiefly a drive around the New Hampshire countryside. Beautiful place to drive, with or without checkpoints. The concours had several spectacular cars, a supercharged K3 Magnette for example, and a couple of incredible restorations, but many entries were simply daily MGs, washed and waxed for the occasion. The gymkhana was more gimmick than go-fast, with a lap in forward gears and a lap in reverse. And the raffle prize was a J2 Midget in dreadful condition; the sort of prize you'd be happy to win and happy not to win.

Mike struck a blow against specialization. His TC placed fourth in class in the concours and was both the only marathon car to win a prize and the only prize winner to run the marathon. Next, he had second-best time of day in the gymkhana. I think there should have been some sort of Big Man of Gathering award, so he could have won it.

On to the banquet. Very clubby, as one would expect and enjoy if one likes MGs, as I do. There were door prizes, one of which was a book, courtesy of R&T. That won me many smiles and appreciations, although in fact I hadn't known about our donation until it was made.

Then they began to announce the marathon winners. Consternation reigned at the marathon table when Bob Pickard won fifth place. Remember, both TCs agreed at the start to waive their handicap. But Bob was eighth on raw time and fifth in the awards, which we took to mean that the waiver hadn't been accepted.

Sure enough, the announcers worked their way higher and higher in the placings, reeling off the names of teams behind us on raw time. Then, the grand prize, first place, let's hear it for Mike and me!

Mike just sat there, half pleased to win and half upset because he had been given the win against his intentions. I loudly whispered that he'd better get up and accept the prize or he'd hurt the feelings of the men who had done all the scoring and set up the rules. His good manners prevailed and he agreed to be the winner.

The next morning Mike started his second 1000 miles of the weekend. Solo. Pleading the press of business, I hitched a ride to Boston in a TF. Along the way we stopped at a garage near the hotel, to deliver parts and advice to a TD that suffered a broken axle in the slalom. That's the way these things go sometimes, he said. Everything went right for me, so I enjoyed the gathering. So did those for whom everything went wrong. I don't know a more convincing recommendation for an event than that. ⊙

Well-wishers and just plain curious thronged the starting line.

The winner, at speed and in reverse.

BACK TO BASICS

*Track testing nine Showroom Stock Sports Cars, the entire field
in SCCA's newest class. Not a fair group, exactly,
but certainly sporting.*

PHOTOS BY JOE RUSZ

THIS COMPARISON TEST is not the ordinary sort of comparison test. In the normal course of events the staff selects a group of comparable cars—economy sedans, say, or sporting GTs on the basis of roughly equal size, price and intent. We put them through a standard series of tests and vote on which car performs normal transportation and entertainment functions best.

We are not going to do that here. This group of cars is a big one and it varies widely indeed when considered on the basis of price or even what the maker intended the car to be. The tests to which the cars were submitted are not the usual tests; some were invented for this occasion. And we did not decide which cars to have in the group.

The Sports Car Club of America did that. Goaded by members who believe that a sports car club should have events for roadgoing sports cars, the SCCA has created a new racing class: Showroom Stock Sports Cars. The sports cars will run with the Showroom Stock Sedans that began racing in SCCA last year and will compete under the same basic rules, their essence being that the cars must have full safety equipment (roll bars, safety harnesses, fire extinguishers, etc) but otherwise be raced exactly as they come from the dealer's showroom. Not even optional wheels or gear ratios are allowed. As an equalization factor all cars may use 165-section radial tires,

any DOT-approved brand but in the diameter that fits the stock wheels be they 13-, 14-, or 15-inch. Having created the class, the SCCA filled it with a selection of models, which must be either 1972 or 1973 model year. The eligible cars are: Fiat 124 Sport Spider; MGB; MGB-GT; MG Midget; Opel GT 1900; VW-Porsche 914 1.7; Triumph GT6; Triumph Spitfire 1500, and VW Karmann Ghia.

Before we discuss inequities or cars which should be on the list but aren't, we must understand that the intent of the class is to have fun, to provide racing for club members in a wide variety of cars. The SCCA doesn't mind one bit that all the major sports car makers have at least one model on the list. And this class is not supposed to prove anything about the merits of the cars.

The original plan was to have a $3000 price ceiling, but that failed because there weren't enough cars on the list. The limit was raised to $4000—with special dispensation for the 914—while the less expensive cars were retained. If there are too many imbalances the SCCA retains the right to delete models. Or if another model comes to mind—say the Saab Sonett—it can be added to the list.

So this is not an equal group. Some of these cars are going to be faster than others. With that in mind, perhaps we should agree now that there will be no proclamation involving the winning car. To say that a modern, expensive sports car with a large engine can outrace an older, cheaper sports car with a smaller engine wouldn't require tests anyway.

What we will look at is the particulars: what each car will do on a race track and why it does it. When we compare cars that way, we'll learn something about both the cars and racing.

One more item about equality. Last August R&T tested nine brands of radial tire. Tires are not the same; a Datsun 240Z would corner at 0.745g on Michelins and 0.713g on Yokohamas, and would stop from 60 mph in 138 ft on Pirellis and 151 ft on Bridgestones. Tires make a difference and there's no sense comparing a Fiat on Michelins to an MG on Dunlops. This is a car test. To eliminate the tire factor, then, we obtained a spare set of wheels for each of the nine cars and fitted each wheel with a Semperit STT M 401 165SR, in 13, 14 or 15-in. diameter. The Semperits performed very well in the August tire tests, so we knew they'd make each car do its best while ensuring that we were testing only the car.

The Tests

Because the cars in the group were picked by the SCCA to be raced against each other, the tests involved here are strictly for track purposes. No consideration was given to daily use; no worry about trunk space, door width or miles per gallon. We wanted to see what the cars would do on the track.

The first checks were preliminary, part of our normal test procedure, measuring acceleration from rest to set speeds and set distances. That's the standard way cars are compared, with 0-60 mph and the standing quarter-mile times and so forth, and we thought we should have these figures for reference.

But those standard figures mean little on a track. SCCA races are begun with rolling starts, usually at 40 mph or more, for safety reasons. A racing car never makes a 0-60 run unless it's coming out of the pits or the weeds. What matters on the track is acceleration out of corners and down the straights.

To measure that, we went to Riverside International Raceway, specifically to the exit of turn 7A. This is a slow right-hander, the second half of an ess, leading onto the back straightaway. The speed out of this turn naturally varies with the cornering capability of the car—a point we'll cover next—but for a stock sports car the average speed past the exit is about 40 mph. Each car was taken through the turn in 2nd gear as fast as it would go and then accelerated through the gears to the markers warning of the end of the straight. Each car was timed from the exit both to a bridge at midpoint and to the markers. We reasoned that this, the amount of time it takes to cover the straight, means more than times from rest to speed.

Cornering power was measured two ways. First was our standard skidpad, timing the cars around a circle of 100-ft radius with the tires set at pressures which our experimentation showed would improve cornering. That's what anybody racing a car would do first, especially if the rules forbid doing anything else.

Next we built a "square circle." This is a racing technique. On the skidpad were placed four corners, each marked with a pylon on the inside and two pylons on the outside, like the gates on a slalom course. Steady-state cornering power is a valid test of a car and certainly useful for comparisons, but driving through a corner is a question of transients. A car must be driven into a turn, around it and out of it, passing from straight ahead to the limit of lateral adhesion and back to straight ahead. The corners were located close together, so that a car's acceleration and brakes wouldn't be a factor: No sooner was a car pointed straight again than it was time for the next turn. Each car was driven around the square as fast as possible and timed. The round skidpad measures a car's steady lateral capability; the square circle measures speed through a real-world corner.

The braking test was also based on the demands of racing. Full stops are not supposed to be required, but braking from high speed to cornering speed in the shortest possible distance is needed. The less distance needed to stop, the more distance that can be covered at speed.

Each car was accelerated to 80 mph, then braked at the maximum usable deceleration (meaning with good steering control) down to 40 mph, with our fifth wheel and electronic counter recording the distance covered while the brakes were on.

And there was a test for practical top speed. There aren't many race tracks with enough space for cars of even this moderate power to reach their true maximums. Some tracks have longer straights than other tracks, though, and here top speed is important; so the nine cars were taken onto Riverside's far corner, turn 8, and given a running start down the long, long hill past the kink and onto the dragstrip, surely as long a run as these cars will ever have in competition.

The final and ultimate test was timed laps around a circuit. By this time we knew which cars were quickest off the mark and to top speed, which cars went round corners best and which stopped quickest. So we laid out a course, using Riverside's turns 7 and 8 and the two straights that connect them. A loop,

FIG. 1. GENERAL SPECIFICATIONS

	Fiat 124 Spider	MGB	MGB GT	MG Midget
List Price	$3867	$3545	$3875	$2699
Curb Weight, lb	2180	2260	2380	1630
Test Weight, lb	2530	2590	2740	1995
Weight Distribution (with driver), front/rear, 90	51/49	51/49	48/52	50/50
Wheelbase, in.	89.8	91.0	91.0	80.0
Track, front/rear	53.0/51.8	49.3/49.3	49.3/49.3	46.3/44.8
Length	156.3	152.7	152.7	137.6
Width	63.5	59.9	59.9	54.9
Height	49.2	49.4	49.4	49.8
ENGINE Displacement, cc	1608	1798	1798	1275
Bhp @ rpm, net	94 @ 5600	79 @ 5350	79 @ 5350	55 @ 5500
Torque @ rpm, lb-ft	94 @ 4000	94 @ 3000	94 @ 3000	67 @ 3250
DRIVE TRAIN Transmission	5-sp manual	4-sp manual	4-sp manual	4-sp manual
Gear Ratios 1st	3.80	3.44	3.44	3.20
2nd	2.18	2.17	2.17	1.92
3rd	1.41	1.38	1.38	1.39
4th	1.00	1.00	1.00	1.00
5th	0.91	—	—	—
Final Drive Ratio	4.10	3.91	3.91	3.90
Layout	front engine/ rear drive	front engine/ rear drive	front engine/ rear drive	front engine/ rear drive
Wheels	13 x 5	14 x 5½	14 x 5½	13 x 4½
Tires (original equipment)	165-13	155-14	165-14	145-13
Front Suspension	Independent, Coil Springs	Independent, Coil Springs	Independent, Coil Springs	Independent, Coil Springs
Rear Suspension	Live Axle, Coil Springs	Live Axle, Leaf Springs	Live Axle, Leaf Springs	Live Axle, Leaf Springs
Brakes	Disc/Disc	Disc/Drum	Disc/Drum	Disc/Drum

mostly, with a fast sweeping turn leading into a decreasing-radius right that exits into a banked left, down the hill into a sharp right, a left and another right, then uphill to the sweeper. All the elements, then, that a car would face on any track.

We also had a variety of drivers: those staff members who either drive in races or have graduated from driver's school or both, plus contributors and former contributors with equal qualifications. One of the latter drivers even wins races.

The Results

BEFORE THE race, though, let's go back to the first of the series of tests. In Fig. 2 are listed the results of the normal times-to-speed and distance tests.

In daily driving as on the track, times to distance mean more than times to speed. In that light the cars can be ranked on the basis of their quarter-mile times and we found that acceler-

FIG. 2. ACCELERATION, STANDING START

	Fiat 124	Porsche 914	MGB	Triumph GT6	Opel GT	MGB GT	Triumph Spitfire	MG Midget	Karmann Ghia
Time to distance, sec:									
0-100 ft	3.5	3.5	4.1	4.3	3.8	4.1	3.8	4.0	4.0
0-500 ft	10.3	10.3	10.5	10.6	10.5	10.7	10.7	10.8	11.3
0-1000 ft	16.1	16.2	16.3	16.9	16.5	16.7	16.8	16.9	17.7
0-1320 ft (¼ mi)	19.3	19.4	19.5	19.6	19.7	20.0	20.2	20.3	21.2
Speed at end of ¼ mi, mph	74.0	73.0	71.0	74.0	72.5	70.0	66.5	68.0	65.5
Time to speed, sec:									
0-30 mph	4.3	4.0	4.6	4.4	3.9	5.0	4.6	4.9	5.3
0-40 mph	6.5	6.3	6.9	6.5	6.2	7.5	7.3	7.5	8.1
0-50 mph	9.4	9.2	10.0	9.3	9.1	10.7	11.2	10.7	12.2
0-60 mph	12.5	12.7	13.7	12.6	13.2	14.6	16.0	15.5	17.5
0-70 mph	16.9	17.5	19.0	17.5	18.3	20.2	22.8	21.8	24.8
0-80 mph	23.0	24.4	27.5	23.4	26.6	29.5	33.3	33.8	39.6
0-90 mph	30.7	34.2		31.6					

Opel GT	Porsche 914/1.7	Triumph GT6	Triumph Spitfire	VW Karmann Ghia
$3346	$4275	$3765	$2895	$2800
2035	2150	2025	1735	1960
2350	2510	2370	2105	2285
55/45	49/51	56/44	53/47	42/58
95.7	96.5	83.0	83.0	94.5
49.4/50.6	52.4/54.0	49.0/51.0	49.0/50.0	51.3/52.7
161.9	159.4	149.0	149.0	165.0
62.2	65.0	58.5	58.5	64.3
47.4	48.4	47.5	47.5	52.0
1897	1679	1998	1493	1584
75 @ 4800	76 @ 4900	79 @ 4900	57 @ 5000	46 @ 4000
92 @ 2800	95 @ 2700	97 @ 2900	73 @ 3000	72 @ 2800
4-sp manual	5-sp manual	4-sp manual	4-sp manual	4-sp manual
3.43	3.09	2.65	3.75	3.78
2.16	1.89	1.78	2.16	2.06
1.37	1.26	1.25	1.39	1.26
1.00	0.93	1.00	1.00	0.93
—	0.71	—	—	—
3.44	4.43	3.27	3.89	3.88
front engine/ rear drive	mid engine/ rear drive	front engine/ rear drive	front engine/ rear drive	rear engine/ rear drive
13 x 5	15 x 4½	13 x 4½	13 x 4½	15 x 4½
165-13	155-13	155-13	5.20-13	6.00-15
Independent, Leaf Spring	Independent, Torsion Bars	Independent, Coil Springs	Independent, Coil Springs	Independent, Torsion Bars
Live Axle, Coil Springs	Independent, Coil Springs	Independent, Leaf Spring	Independent, Leaf Spring	Independent, Torsion Bars
Disc/Drum	Disc/Disc	Disc/Drum	Disc/Drum	Disc/Drum

ation from rest is a function of power-to-weight modified by gearing: we have the Fiat, the most powerful car in the class, crossing the line first; then come the 914, the MGB, the GT6, the Opel, etc. The GT6's relatively tall final drive gears put it behind the MGB despite the MG's greater weight. The gap is not as wide as the spread in power-to-weight would lead one to expect, however. The Ghia has half the power of the Fiat and is the slowest of the nine, but even so it lags by only 2 sec at the finish. On the basis of a normal road test, we'd have to say that the difference in performance is not extreme.

But let's move to the track. In Fig. 3 we have the times from turn 7A to the shut-off markers at the end of the straight. There are three times for each car; the total from corner to marker, from the corner to the bridge, and from the bridge to the marker. Interestingly, the order has changed. The lightweight Opel has moved from fifth place at the drags to first past the marker. The GT6 is as close behind the Opel here as the Opel was behind the GT6 from a standing start. And the Fiat has moved back to third. Yet the power and weight hasn't changed at all. But the use of gears and engine speeds for this test is different. The Opel has a tall final drive. Note that it's especially quick from the corner to the bridge; it came through the corner at peak torque in second gear, with plenty of space left on the tachometer, and was peaking in third as it reached the

FIG. 3. RACE ACCELERATION
Times from Corner 7A to Two Points
In Seconds

	Corner to Bridge	Bridge to Marker	Corner to Marker
Opel GT	13.2	4.9	18.1
Triumph GT6	13.4	4.8	18.2
Fiat 124	13.6	4.9	18.5
MGB	13.8	5.0	18.8
MGB GT	13.9	4.9	18.8
Porsche 914	13.9	5.2	19.1
Triumph Spitfire	14.5	5.1	19.6
MG Midget	14.5	5.4	19.9
VW Karmann Ghia	14.6	9.5	24.1

FIG. 4. LATERAL ACCELERATION Round Skidpad		
	Pressure, F/R, psi	Lateral Accel, g
MGB	36/30	0.795
Fiat 124	36/32	0.783
Porsche 914	30/34	0.758
Triumph Spitfire	28/32	0.758
Opel GT	32/32	0.758
MG Midget	30/36	0.758
MGB GT	36/30	0.745
Triumph GT6	34/32	0.745
Karmann Ghia	36/36	0.713

FIG. 5. SQUARE SKIDPAD Lap Times	
	Time, sec
MGB	13.1
MGB GT	13.1
Porsche 914	13.2
MG Midget	13.5
Opel GT	13.7
Karmann Ghia	13.7
Triumph GT6	13.8
Triumph Spitfire	13.8
Fiat 124	14.0

FIG. 6. BRAKING DISTANCE 80-40 mph	
	Braking distance, ft
Porsche 914	239
Opel GT	270
MGB GT	278
Fiat 124	280
MGB	296
Triumph GT6	305
MG Midget	306
Triumph Spitfire	307
Karmann Ghia	325

shut-off marker. Thus it didn't have to be dropped into a relatively dead top gear for the stretch.

Same for the GT6, except that its gearing is taller still, meaning that the GT6 is still pulling strongly when it goes past the marker—so strongly that it nearly closes the gap on the Opel.

This works both ways. The Fiat is wound tight in second at the corner and uses 3rd and 4th gears for this distance. The MGB only needs 2nd and 3rd but it can't match the Fiat's power, the MGB-GT is heavier still, and so on. The 914 particularly suffers here as it comes with an overdrive 5th for the open road and a very low first gear for city starts. From a standing start it gets a great leap forward and a good elapsed time for the quarter mile. But once into the upper gears, the lack of power and engine speed hurts it.

And the Ghia? Pure lack of power. For the first segment of this test it's merely slow, lagging behind the others but not by much. At the bridge, though, the Ghia is shifted into top gear and acceleration becomes merely perceptible as the gap widens.

What all this means is that if all nine cars came through turn 7A in a pack three wide and three deep, by the end of the straight they would no longer be a pack. They would be a string, with 6 sec between first and last.

But, you may think here, what about cornering power? Surely these cars must vary in cornering power and this must influence the speed down the straight. Faster in, faster out, so to speak.

Yes. But. In Fig. 4 we have the skidpad figures, taken (as mentioned) with tire pressures adjusted to give each car its maximum grip. Our first finding in this test was that the cars weren't all that far apart, not the way one would expect.

We knew from driving the cars on the road that the MGs were harshly sprung, that the Fiat was soft and comfortable, that the 914 seemed both firm and comfortable. We would have projected from theory and racing practice that the 914 would have been very good. And if we listened to folklore we would have expected the Triumphs with their swing axles and the Ghia with its leaning front suspension to be both tricky and slow.

But it didn't work out this way. The MGB—live axle, leaf springs, coal-cart suspension and all—went around the fastest. Then came the soft Fiat. The sophisticated 914 tied with the Spitfire, Midget and Opel. The extra weight of the MGB-GT counts against it and the Ghia is at the back again. In its favor, though, was the security with which the Ghia reached its maximum. The rear wheels jacked up a bit but the old-style front suspension ensures that Ghia's steady-state cornering is done in an attitude of understeer, even to the inside front wheel lifting off the ground.

As an adjunctive lesson, we learned that some cars can gain more from adjusted tire pressures than others. The Fiat, the MGB-GT, the MGB and the 914 gained the most. But the Midget gained only a tiny bit and the Spitfire didn't gain at all. More pressure simply made it skitter over the high spots.

But just driving around corners isn't our main interest here. In the squared circle, remember, we put each car into, through and out of four turns.

The results of that are in Fig. 5. For once we don't have a new name at the top of the list: The MGB was best here, as it was going around the normal circle. But the MGB-GT tied with it. That's a surprise, the explanation for which is buried 'way back in the specification table. The MGB-GT has more weight on the rear wheels, is in fact tail-heavy whereas the MGB has a forward weight bias. In this test, then, a car which can be coaxed into or naturally tends to oversteer does well. In a tight turn, the understeering car must be forced around, front wheels scrubbing off speed. The oversteering or neutral car can be pointed, wagged and straightened out. The 914 has gained on cars which it formerly only equaled. And look at the Ghia—better than both Triumphs and the Fiat! The latter car looked especially poor in this test. It understeers and rolls and does not do well in sharp changes of direction, despite its good showing on the normal circle.

The 914 has the mid-engine's low polar moment of inertia as well as what we judged to be the best and most precise steering. With enough power at a high enough speed, the tail can be provoked into a slide. And the front end darts toward

the inside of a turn the instant the throttle is closed. So the car could be hurled about with great gusto. We let it drift wide under power into the turn, whipping it sideways by lifting off and flinging the tail around with more power as the inside front wheel bounced off the ground. It all felt great and looked spectacular, but all that flamboyant flinging about only scrubbed off speed and tire tread. The best times for the 914 came with the car driven carefully and tidily.

The Opel displayed a chronic problem: in a tight turn the inside rear wheel lifts, the engine races, the tire spins and the car just sits there until the speed and the wheel both drop. The Opel didn't come out of the turns fast, then, because it couldn't power out.

Only in the braking test did the cars perform according to expectations. Fig. 6 shows the 914 best, slowing from speed in the shortest distance. That's what you'd expect from four disc brakes and a slight rearward weight bias.

The Opel, MGB-GT and Fiat were as easy to control as the 914 was but didn't have the stopping power. The MGB's weight distribution hurt it more than its lighter weight helped it. The GT6 swerved under maximum pressure and the Spitfire and Ghia both tended to lock up wheels if pushed.

And we discovered an interesting thing about which perhaps we shouldn't tell you. The Fiat has vacuum assisted brakes, too much assisted in our opinion. This makes controlling the brakes difficult, so one of our guest drivers popped under the hood and pinched the vacuum line shut. What the SCCA technical inspectors would say about this we don't know. But it did make the Fiat's braking easier to modulate.

This test had the widest variation among the cars. If we can assume that the cars are all bunched up on the straight, heading fast for a slow corner, then the car which needs the longest distance to slow down must be the first one to get off the power and onto the brakes. From the bunch, then, the Ghia drops out first, then the Spitfire, then the Midget, and so on.

This is compounded in that the cars don't slow down from the same speed. As we've seen from the acceleration figures, they won't reach top speed at the same time. And as we see in Fig. 7, the top speeds reached by a downhill run from the far corner of the track vary from car to car. The Fiat, for example, hit 5500 rpm in 5th, below the redline. And the 914 never needed 5th at all. At the other end, the Spitfire, Midget and Ghia are going as fast as they can go, albeit not very fast.

Now we have all the measured data on acceleration, handling and braking. We know which cars are the quickest from rest and from slow corners, which cars go around corners and through turns better than their competition, which cars stop shortest and which go fastest on the long straight. For a bringing-together sort of test, we wanted to find out just how important each of these factors is: is cornering power as important as engine power, does good braking make up for lack of speed?

So we arrive at Fig. 8, the lap times for our road course. The times shown are the result of many laps by several drivers, each one of whom was doing his best in whatever car he had at the time. These figures don't quite speak for themselves.

FIG. 7. TOP SPEED At Riverside	
	mph @ rpm
Fiat 124	101 @ 5500
Opel GT	98 @ 5000
Triumph GT6	98 @ 4800
Porsche 914	96 @ 4100
MGB GT	96 @ 5300
MGB	94 @ 5200
Karmann Ghia	92 @ 4350
Triumph Spitfire	86 @ 5100
MG Midget	85 @ 5200

FIG. 8. LAP TIMES	
	Time, sec
Triumph GT6	54.4
Opel GT	54.9
Porsche 914	55.2
MGB	55.2
Fiat 124	55.4
MGB GT	55.5
Triumph Spitfire	56.3
MG Midget	57.3
Karmann Ghia	58.3

All they tell you is which cars went around the circuit faster than which others. Why and how come from subjective assessments from the drivers, given in the order of increasing lap times.

Triumph GT6—does everything well. Plenty of power, gearing that lets the power be used in these conditions and good brakes. The car's natural understeer brings it around the fast sections under control, and the power lets the driver fling the car around the tight turns.

Opel GT—has power, like the Opel sedans that overwhelmed the Showroom Stock class last year. And it corners and brakes fairly well. What hurt the Opel GT was the lifting inside wheel. The engine couldn't be used fully until the car was pointed straight ahead and cruising through the turns in 3rd gear was as quick as sliding and spinning in 2nd.

VW-Porsche 914—is listed first in a tie with the MGB because the fastest driver in the group said that if he just had enough time to figure out how the 914 would go best he could beat the MGB *and* the GT6 and Opel. You don't get to the winner's circle saying "if" but if the handling had been more flexible or easier to learn, then maybe the 914 could have made up for its lack of power. The 914 was a bit of a disappointment.

MGB—has more than tradition going for it. It feels harsh and clumsy but doesn't look that way on the clocks. The MGB's secret of success is that a driver can use everything the car has.

Fiat 124—is perhaps the best of the boulevard racers. It has a lovely engine, a lot of power and soft suspension. It understeers, but the Fiat was in fact the only car to spin during the test. We learned that the Fiat went best when driven gently around the fast corners and wound to redline on the straights.

MGB-GT—take everything said about the MGB and add weight. The good behavior through tight turns is caused by the same heavy tail that makes the coupe a handful in the fast turns. Now we know why all the MGBs in production racing are convertibles.

Spitfire—lap times aren't everything. The Spitfire is better balanced than the GT6, and it's more fun. Excellent throttle response, so while the car's attitude varies with power, the power can be properly controlled. The Spitfire just lacks speed. It's deceptive, for though the Fiat felt slower than it was, when our drivers climbed out of the Spitfire they went right to the timers expecting to have set a record or two. They didn't, but they enjoyed trying.

MG Midget—won nobody's heart. The lack of speed was expected but the stiff and vague steering, the roll oversteer, the mushy front end and the general feeling of cramped obsolescence were things we had forgotten. The Midget won't win races because it doesn't go fast.

Karmann Ghia—even slower than the Midget, but at least more fun. Or maybe it's just the underdog aspect. Either way the Ghia has a good ride and proper steering, and it can be driven on the ragged edge without incident and with drama: what fun to watch it lift its inside front wheel just like a Porsche 911S! The Ghia is out of contention here. It won't win, but neither will it be disgraced.

What have we proved? In terms of values, in deciding which performance factors are most important on the track, the deciding factor seems to be power. Power alone won't do it: witness the Fiat. But when we check the results and find that the two quickest cars down the straightaway finished one-two in lap times, with the best-handling MGB and best-braking 914 tied for third place, we are forced to conclude that getting from corner to corner quickest will do more for a car than getting through corners will.

Well, what of it? We already knew that the SCCA organizes production sports car racing to provide an equal chance for different makes and that Showroom Stock Sports Car racing is supposed to be for the man who wants to race his car, win or not.

For those who must win in Showroom Stock, the GT6 is most likely to do it. For those who can afford a dual-purpose car and like to be comfortable going to and from the track, the Fiat or the 914. For fun, the Spitfire. And to prove that VW does make a sports car, the Ghia.

PHOTOS FROM PAUL FRÈRE

Driving Impression:
MG BGT V-8

*MG fans would love it, but
it won't be available in the U.S.*

BY PAUL FRÈRE

F OR THE TIME being the V-8 version of the MG BGT is
on sale on the British home market only, but however
long you wait you apparently will never be able to buy
one in the United States: British Leyland has decided it would
be too expensive to make it comply with the innumerable rules
it would be faced with before getting into U.S. dealers'
showrooms. The American MG enthusiast is definitely the
poorer for it, because with an engine of twice the capacity of
the original unit under its hood the V-8 is as much an MG
as any MGB ever made while being infinitely better suited
to American driving conditions and habits than the 4-cylinder
version. In fact, it has direct American ancestry, for the aluminum V-8 unit it has inherited from the Rover 3500—another
British Leyland product—started life in 1961 powering the orig-

inal compact Buick Special, Olds F-85 and Pontiac Tempest.
   The version used in the MG is the latest "detoxed" one with
its compression ratio lowered to 8.25:1 and an entirely new
intake manifold aspirating from twin constant-depression
Strombergs located at the rear end of the engine, so that not
even a "power bulge" is required to squeeze the V-8 under
the standard hood. Though it has twice the capacity of the
four it does not deliver anything like twice its power; but its
137 bhp DIN (about 131 SAE net) is enough to raise the MG's
performance into an altogether different class. And there is

no penalty in handling; the alloy V-8 is actually *lighter* than the four, though its heavier ancillaries end up slightly increasing the weight hidden under the hood. Uprated brakes are also responsible for the slightly higher overall weight of the V-8 model.

Other modifications are strictly related to the more powerful engine: they include a slight relocation of the rack-and-pinion steering box and a universal-jointed column to clear the engine, higher intermediate gear ratios, a standard overdrive, and stiffer rear springs—mainly to prevent axle windup with the increased torque, I suppose. The diameter of the front anti-roll bar has also been increased, presumably to maintain the front-to-rear roll-stiffness balance. Wheels with cast alloy centers and steel rims and a heated rear window are standard items on the V-8.

No attempt has been made to otherwise update the basic model, which externally is still very pretty and quite modern but certainly could do with a better dashboard arrangement and less haphazard controls for the heating and ventilating system. I was happy to drive the car on a reasonably warm day, for it proved impossible to get any heat worth mentioning from the heater, though it may only have needed bleeding. Personally I also like a proper 5th gear much better than an electrically controlled overdrive. Not only is it more reliable and entails a much smaller weight penalty, but it also seems quite illogical to me to control 5th gear in a way different from the other four. And if I must contend with an overdrive, then the switch should be in the gear-lever knob so that the gear lever and the overdrive switch can be manipulated simultaneously: an overdrive stalk mounted on the steering column to be operated with the same hand as the gear lever is certainly the most absurd setup imaginable. And that's what the B has.

Even if nobody seems to take any notice of the overall 50-mph speed limit introduced in England because of the alleged fuel shortage, it was not possible for me to take any performance figures. But even though the V-8 is obviously a much better-than-average performer I was rather surprised when I went through the *Motor* road test and saw the car would do 0–60 in 7.7 seconds, 0–100 in 23.4 sec and the standing quarter-mile in 15.8 sec, and a mean maximum speed of fractionally over 125 mph. Subjectively the car feels slower than this, probably because the subdued V-8 burble makes the engine appear so lazy and also because the apparently quite heavy flywheel defeats spectacular fireworks in the lower gears. And as is often the case with a big, lazy engine in a comparatively light car, within certain limits it does not seem to make much difference which gear is selected. So it seems rather wasteful to have five ratios to choose from, though the gearchange itself with its short, precise movements and quick, unbeatable synchromesh deserves full marks.

Though the ride is quite acceptable when the car is driven fast on good roads, indifferent country lanes show up the "vintage" character of the suspension and the ride becomes harsh. One wonders how much more it would have cost to adapt Jaguar E-Type rear suspension to the MG V-8. Handling and roadholding have not suffered, however, and though this basically 12-year-old design running on 5-inch-wide rims cannot be expected to offer the ultimate in cornering power, handling is nice and predictable, only slightly biased toward understeer but with plenty of power to hang the tail out if required. These good manners are enhanced by the accurate steering, which provides good feel and has (at last) exchanged its oversize steering wheel for a more reasonably sized, soft-rimmed one. Good feel is also provided by the servo-assisted brakes and heel-and-toeing is easy, but I had no chance to find out how hard the brakes can be used before they start to fade. I also liked the new pleated brushed nylon front seats.

The engine installation is quite neat and important items are much more accessible than the shoehorning operation might suggest. Only the plugs are rather awkward to get at, and the hydraulic tappets make the removal of the cast aluminum valve covers unnecessary.

For the rest, the car is known well enough to make further

descriptions unnecessary, though it should perhaps be recalled that the "+2" accommodation is strictly for small children over short distances and that the car should be considered essentially a 2-seater with ample luggage accommodation, easily accessible through the large lift-up rear door.

Despite its shortcomings and the typical British Leyland failure to update a car that is certainly still worth the effort, I could certainly live with this MG. It reminds me very much of the late lamented Ford V-8 engined Sunbeam Tiger, of which I accepted the obvious shortcomings in exchange for its endearing virtues. The main trouble with the MG is that even in its home market the same money (over $5500 in Britain) would almost buy a 3-liter Capri (nearly equaling its performance) *and* a Mini for shopping.

SPECIFICATIONS

ENGINE

Type	ohv V-8
Bore x stroke, mm	88.9 x 71.1
Displacement, cc/cu in.	3528/215
Compression ratio	8.25:1
Bhp @ rpm, SAE net	131 @ 5000
Torque @ rpm, lb-ft	185 @ 2900
Fuel requirement	94-oct

DRIVETRAIN

Transmission	4-sp manual + OD
Gear ratios: OD (0.820)	2.52:1
4th (1.00)	3.07:1
3rd (1.26)	3.86:1
2nd (1.97)	6.06:1
1st (3.14)	9.63:1
Final drive ratio	3.07:1

CHASSIS & BODY

Body/frame	unit steel
Brake system	10.7-in. disc front, 10.0 x 1.7-in. drum rear; vacuum assisted
Wheels	alloy center, steel rim; 14 x 5J
Tires	Dunlop radial, 175HR-14
Steering type	rack & pinion
Turns lock-to-lock	2.9
Suspension, front/rear: unequal A-arms, coil springs, lever shocks, anti-roll bar/live axle, leaf springs, lever shocks	

GENERAL

Curb weight, lb	2425
Wheelbase, in.	91.1
Track, front/rear	49.0/49.2
Length	154.8
Width	60.0
Height	50.0
Fuel capacity, U.S. gal.	14.0

MGB, 1962-1967

A lot of fun for very little money

"THE SPORTS CAR connoisseur will find in this latest MG challenger all that he has been looking for . . ." Those are the opening words of the British Motor Corporation's (now British Leyland) sales brochure for the MGB when it was introduced in late 1962. The B was the latest in a long line of sports cars from Morris Garages, the original firm named for the founder, Sir William R. Morris. It succeeded the MGA which had been in production for seven years and established a sales record never previously equaled by any sports car: 100,000 cars.

The MG octagon made its initial appearance in 1923 when a Morris Oxford chassis of the period was fitted with a Hotchkiss engine, and went out into the world of competition. MG was a name to be reckoned with in automobile racing until 1935 when the program was allowed to terminate because the management felt that it was no longer achieving sufficient research and development information from the racing program.

MGs arrived in the United States after World War II as the forerunners of an invasion of sports cars from abroad that has continued and grown to very large proportions. In the late 1940s and early '50s, the sight of an MG TC, and the slightly more modern looking TD model, was enough to stir the heart of many red-blooded American youngsters who dreamed of someday owning one. It seems to us that much of the affection

we still feel for MG cars is the result of those early days of boyish (and girlish) dreams of traveling briskly down our favorite country road behind the wheel of a TC or TD with the top down and the wind whistling over, around and through us.

The MG TD was followed by the short-lived MG TF model with a larger engine and more rakish lines, but still retaining the original and what we thought of as traditional MG styling. In 1955, however, the old world gave up the struggle and the MGA was born with its more modern styling based on various racing and experimental car designs and using the B series engine with a displacement of 1489 cc. In 1959 the engine displacement grew to 1588 cc, which lasted until 1961 when it was again increased, this time to 1622 cc. The MGA was produced until June 1962 when it gave way to the MGB.

The MGB was an almost totally new design inside and out. Not only was the styling a departure, but perhaps the biggest news was that the traditional ladder-type frame had given way to monocoque construction.

The engine was a continuation of the B-series engine but further expanded to 1798 cc by enlarging the cylinder bores but with no increase in stroke. The crankshaft and block casting were strengthened to handle the increased displacement and power output. Horsepower was now 94 bhp @ 5500 rpm and the compression ratio dropped from 9.4:1 to 8.75:1.

Although the B featured a new Borg & Beck diaphragm-spring clutch for lighter effort, it continued to use the A gearbox with its non-synchromesh 1st gear. *Road & Track*, along with other automotive journals, lamented the lack of an all-synchro gearbox but it was not until the 1968 model that this became available, a year after the models covered in this Used Car Classic.

The final drive ratio for the MGB was 3.91:1 rather than 4.10:1 as in the A but there was also a change to 14-in. wheels from 15, so there was little difference with the higher final drive. Acceleration was still brisk and performance good. Overdrive was an extra-cost option that many MGB owners wished they had ordered after driving their cars awhile and discovering that they were churning along at 3340 rpm at 60 mph.

In terms of accommodation, the B was a vast improvement over the A with considerably more hip room, more comfortable seats with seatback adjustment and gobs of legroom. Headroom in the roadster was also quite good and most road tests and reports on the car commented on how good the visibility was even with the top up because of the large plastic rear and quarter windows.

In October 1965, BMC introduced the MG BGT, a closed two-seater with a spacious coupe body and easy access from the rear through a large liftback. The BGT was essentially the same as the roadster in its mechanical details with the exception of stiffer springs front and rear to take care of anticipated heavier loads and the addition of an anti-roll bar in front. MG felt that the BGT would fill a gap in the market for those who wanted the performance of a sports car with the comfort and security of a closed car.

Both the B roadster and the GT are relatively simple and straightforward cars, a large portion of their charm. The suspension is nothing fancy, featuring a live rear axle with leaf springs and A-arms with coil springs up front. The drivetrain is neither unusual nor exotic. What these cars offer is driving pleasure for the enthusiast along with a reasonable degree of reliability and, with the roadster especially, a lot of fun.

What to Look For

THE FIRST decision facing the MGB Used Car Classic buyer is rather an obvious one: roadster or GT. The former offers much of the historical MG flavor of driving: top down, wind in the face, sky above and hassles with the top. The latter is snug and warm no matter what the weather and gives the owner extra carrying capacity for parcels and luggage but not people.

As with many British cars, the MG is susceptible to rust problems. The common areas to check thoroughly are the rocker

The BGT version was introduced in 1965 to add another dimension to the MGB series. Mechanically, it was nearly identical to the roadster.

panels, above the rear wheel openings, along the seams on the top of the rear-quarter panels and front fenders, around the front turn-signal and parking-lamp assembly and the floor area of the cockpit.

Many MGs will show signs of the infamous "MG crack." This is a quite noticeable crack on the doors starting at the rear post of the vent windows and running down toward the bottom of the door, usually extending about six inches. It is not just a crack in the paint but in the metal itself in many cases, and it is more prevalent in the roadsters than in the GT models.

The convertible top comes in two configurations: the traditional stow-away is detachable from the erector-set-like supports so that the whole apparatus can be removed and put into the trunk; the fold-down top folds into the luggage compartment behind the seats.

Both types suffer from improper folding and a car of this vintage (1962–1967), if it still has the original top, will likely require a new one. The plastic windows have a tendency to discolor and crack with age, seriously inhibiting rearward visibility.

The B-series engine is quite robust. It was strengthened in 1965 and later models with the change from three main bearings to five, so you may want to consider a 1965 or later car. All MGBs delivered to the U.S. were equipped with external oil coolers mounted in front of the radiator. A weak point of the engine, especially the five-bearing model, is the head gasket between the number three and four cylinders. If a head gasket is going to go, it will probably be in that area. You should always have a compression check performed on the car before agreeing to purchase it.

Another problem that crops up on MGBs is cracked cylinder heads resulting from overexuberant mechanics applying too much torque on the bolts or following an improper tightening sequence, so this is also a must for examination. Valve noise is common and is a result of loose settings for rocker clearance given by the factory.

Getting the SU carburetors to work together is of vital importance in terms of the MGB's performance. Many MG owners have tried different jet sizes but most have come back to the factory stock size as it really works best. Balancing and coordinating the operation of the two carburetors is the best answer for performance and economy, not modification.

In the area of emission controls, the pre-1968 MGBs had only the PCV to contend with, which is why we have cut off the model years at that point for this report, but proper maintenance of the PCV is very important. If the rubber diaphragm inside the valve is the least bit brittle or cracked, it will pull oil from the sump and pass it through the induction system, producing a large amount of smoke from the exhaust. If in doubt, change the valve and do not pull the head off to do a valve job without checking the PCV first.

Pre-1968 cars without synchromesh may have damaged 1st gears, and this should be checked out. Also, you'll often find that the synchro in 2nd gear has deteriorated or disappeared. Many enthusiasts do not mind this overly much and feel it gives them good practice at double-clutching. If you do have the problem and wish to replace the synchro be sure to discuss it thoroughly with the service manager at an authorized dealer as there is a replacement synchro which is much better than the original unit.

Overdrive was available as an option on the MGB models covered in this report and, if you can find one it is well worth having. It makes the B a much more pleasant highway cruiser. However, there were not very many of the U.S. models which had the option and if you find one that does, make sure it's operating properly because repairing it is expensive.

MG suspension design is not the most modern in the world but it does get the job done. In examining a used car, check the condition of the lever-arm shock absorbers by pushing down on each corner of the car. If the car bounces quite a bit you can be certain the shocks are badly worn.

MGB ROADSTER & BGT
MODEL DESCRIPTIONS & CHANGES

1962: MGB roadster introduced to the U.S. market; all-new body & chassis, 1798-cc, 94-bhp net engine. Replaced MGA, had more room for driver and passenger, greater power output and improved handling.

1965: Engine strengthened by conversion from three to five main bearings. October 1965 the BGT introduced featuring a closed car with greater carrying capacity and large rear door opening. Mechanical components identical with the B roadster; weight difference 230 lb.

PERFORMANCE DATA
From Comtemporary Tests

	1962 B Roadster	1966 BGT
0–60 mph, sec	12.5	13.6
0–90 mph, sec	34.5	37.2
Standing ¼ mi, sec	18.5	19.6
Avg fuel economy, mpg	26.5	23.0
Road test date*	11–62	5–66

*A set of MG road-test reprints is available from R&T's Reader Service Dept for $5.00 plus 50¢ postage per order. The set includes 24 tests from the 1949 MG TC to the 1971 MG BGT.

BRIEF SPECIFICATIONS

	1962 B Roadster	1966 BGT
Curb weight, lb	2080	2308
Wheelbase, in.	91.0	91.0
Track, f/r	49.2/	49.2/
	49.2	49.2
Length	153.2	153.2
Width	59.9	59.9
Height	49.4	49.8
Fuel capacity, gal.	12.0	12.0
Engine type	ohv inline 4	
Bore x stroke, mm	80.3 x 89.0	
Displacement, cc	1798	1798
Compression ratio	8.75:1	8.8:1
Bhp @ rpm, net	94 @ 5500	98 @ 5400
Torque @ rpm	107 @ 3500	107 @ 3500
Gearbox	4-speed, non-synchro 1st gear	
Final drive ratio	3.91:1	

Look at the shock bodies too and if they show signs of leaking they are no good and should be replaced, no matter what the owner may tell you. Lever-arm shocks cannot be disassembled and repaired, they must be replaced. Koni makes a kit for the rear suspension that adapts regular tube-type shock absorbers to the B. The potential buyer should also check the wheel bearings and the condition of the rear leaf springs. These springs have been known to break periodically and you should be aware of their condition before you purchase a car.

MGBs, like their predecessor MGs, came with a choice of wire or disc wheels. Most enthusiasts know the perils of wire wheels and the cost of keeping them in good shape and will recommend going for the discs. Also, if you are planning any sort of competition with the car, such as slaloms or gymkhanas, disc wheels are stronger and safer. If you must have the wire wheels or the car you want has them, they should be checked by someone who knows about wire wheels, spokes should be replaced if they are bent or broken and the application of a little paint can make them as good as new.

MGs generally tend to run cool and they will start quickly as long as the temperature is no lower than 0 degrees F. The oil pressure should register 55–60 psi or thereabouts, fluctuating somewhat at a stop.

All things considered, the MGB is probably an excellent choice in these times. It offers relatively economical driving with a minimum of fuss over tune-ups and valve adjustments, straightforward handling characteristics and reasonable performance. It has been with us for a sufficiently long period of time that replacement parts are easily attainable and not terribly expensive. The motoring enthusiasts who are looking for an inexpensive form of entertaining driving should find the MGB fills their needs quite nicely.

TYPICAL ASKING PRICES

Year & Type	Price Range
1962–64 B roadster	$500–800
1965 roadster & BGT	$700–1000
1966 roadster & BGT	$800–1200
1967 roadster & BGT	$900–1400

Prices highest in the northeast and west, lowest in midwest and south.

PHOTO BY JOE RUSZ

DRIVING IMPRESSION:
1967 MGB ROADSTER

WE WANTED TO drive a 1967 MGB as '67 was the last year MGBs were made without an over-abundance of emission-control equipment. We were aided in our search for a suitable car by the Long Beach MG Club of Southern California, one of the largest and most active MG clubs in the U.S. They put us in touch with member Joe Heinz, who is the proud possessor of a '67 B he purchased new in Europe.

Joe came to our offices in Newport Beach one gorgeous afternoon and we took his car out for a spin down Pacific Coast Highway, running along with the top down, the sun warming us and the wind providing just the right amount of exhilaration.

The car has accumulated slightly more than 94,000 miles over the years and has won several concours d'elegance. Thus far Heinz has avoided having any major work done and his service records show such relatively minor entries as a new head gasket in June 1970, rebuilding the brake master cylinder in November 1971 and so on. Joe is a strong believer in preventive maintenance and has had his car serviced about every 2000 miles throughout his tenure as owner.

Having disposed of a 1968 MGB roadster little more than a year ago, I found the Heinz car quite familiar and comfortable upon entry, with all the switches, gauges and controls in the proper places. The engine fired up immediately on the first turn of the starter motor and belied its 94,000 miles of operation, running smoothly and quietly.

Even for my portly, over-six-foot frame, the B is quite spacious and there is an abundance of legroom. The pedals are well situated for heel-and-toe down-shifting and the shift lever is correctly located with relatively short throws. The car moves out quite briskly and revs freely up to about 4500 rpm where it begins to feel a bit strained. Joe's B roadster is not equipped with the optional overdrive unit; maintaining a high cruising speed for any length of time can be a rather noisy and tiring experience.

The handling and cornering characteristics of the B are satisfyingly normal and predictable and it's quite an easy car for the novice sports-car driver to manage. I have long maintained that an MG is a near ideal choice for the first-time sports car buyer because it is relatively simple and straightforward, yet offers a good deal of motoring fun for the dollar. The suspension is on the firm side and there is a stiff-legged feel to the car on bumpy and irregular surfaces.

My memories of the MGB with the top in position are that it is one of the coziest cars imaginable, especially on a rainy night. Certainly the convertible top is not entirely free from leaks, but it does a good job of keeping wind and rain out and does not encroach much on usable headroom except during entry and exit. The Heinz car has the top which folds down into the boot rather than the totally removable type; however, with the rollbar the top will not fold down properly so Joe has to remove it entirely which is quite an operation.

It was great fun to be at the controls of an MGB again and gave rise to feelings of regret at having sold mine. Although it is not an especially sophisticated or exotic automobile, it does present the driver with a good deal of pleasurable driving, which is more than can be said about many cars.

—*Thos L. Bryant*

ROAD & TRACK
R&T
ROAD TEST

MG MIDGET MARK IV

It's showing its age but the basic elements of
driving pleasure are still there

PHOTOS BY JOE RUSZ

ANYTIME THERE IS a discussion of sports cars, the name MG must come to the forefront. In the late Forties when the sports car movement first came to America, it was the MG TC that led the way and set the parameters of what a sports car should be. Almost 30 years later, MG may be sharing the market with a lot of other marques, but the basic sports car in this country has to be the MG Midget.

This diminutive roadster is the smallest and least expensive sports car available in the U.S. at this moment. The Midget has a long history going back to 1958 and the birth of the Austin-Healey Sprite, that odd-looking but lovable Bugeye that

represented a return to the basics. Over the past 18 years, the car has been updated several times, starting with the Sprite II in 1961 and the emergence of the MG Midget nameplate that same year, until today what we have is a car with roll-up windows, carpeting and relatively luxurious interior features compared to the original concept. (The Sprite name disappeared in 1970.) The engine has grown from the original 948 cc to 1493, the power from 45 net bhp to a high of 65 (with the 1275-cc engine) and now to 55.5 bhp (50 in California). And the price? It's gone from $1795 to almost $4000.

Despite the evolutionary changes that have taken place, the

61

Midget is, by any standards, an old-fashioned car. However, this quality leads to a simplicity that is rare today and helps keep the price at the bottom of the sports car scale, albeit that scale has taken a quantitative leap upward in recent years. However much automotive journalists and critics may call for a new design from British Leyland to replace the present Midget, it's unlikely that will happen unless or until there is a direct competitor for the car and it can no longer meet U.S. safety and emission standards through alteration rather than remaking.

Under the BL scheme of consolidation, the Midget now shares the same engine as the Triumph Spitfire: an overhead valve, inline 4-cylinder. With the displacement increase from 1275 to 1493 cc, the Midget has taken on a more relaxed attitude. It's really no quicker up to speed thanks to emission controls, but it doesn't work quite so hard and there is a little less of the busy, noisy effort of earlier Midgets. The larger engine's better low- and middle-rpm torque range accounts for the difference. Once the engine is taken over about 4000 rpm, however, it becomes a typically buzzy 4-banger. The 4-speed, fully synchronized transmission mates well to the engine and is a pleasure to use. Shifts can be accomplished crisply and with a relatively short throw, but we lament that there is no optional overdrive with the Midget as there is with the MGB

and Spitfire.

The Midget sits on an 80.0-in. wheelbase (11 in. shorter than the MGB) and has an overall length of 141.0 in. (18 in. shorter than the B and about the same length as the sub-compact Renault 5). Suspension is by lower A-arms, lever shocks as upper lateral arms, coil springs and anti-roll bar in front, with a very simple arrangement of live axle on leaf springs and lever shocks at the rear. The result is a ride that is firm and jouncy in the best tradition of British sports cars and with handling that is best characterized as a bit twitchy. The rack-and-pinion steering requires only 2.3 turns, lock-to-lock, and this gives the Midget a maneuverability factor that's hard to match. But the inherent understeer can change suddenly to final oversteer if the car is pushed hard enough. The first time out driving on the highway can be rather exhilarating if you're not prepared for all of this as the Midget will change direction quicker than you can say "God save the Queen!"

The brakes could stand improvement. After four ½g stops from 60 mph, the brakes had faded completely away. We repeated this test twice with ample time in between for cooling and recovery with no improvement. The brakes weren't very effective in our stopping distance tests either; the front wheels locked easily, causing the Midget to dart about. So the only thing to do was to use the pedal gently and pay the penalty

PRICE	
List price, all POE	$3949
Price as tested	$4219

GENERAL

Curb weight, lb	1775
Weight distribution (with driver), front/rear, %	52/48
Wheelbase, in.	80.0
Track, front/rear	46.3/44.8
Length	141.0
Width	54.0
Height	48.3
Fuel capacity, U.S. gal.	7.5

CHASSIS & BODY

Body/frame	unit steel
Brake system	8.3-in. discs front, 7.0 x 1.1-in. drums rear
Wheels	styled steel, 13 x 4½
Tires	Pirelli Cinturato CF67, 145 SR-13
Steering type	rack & pinion
Turns, lock-to-lock	2.3
Suspension, front/rear: lower A-arms, lever shocks as upper lateral arms, coil springs, anti-roll bar/ live axle on leaf springs, lever shocks	

ENGINE & DRIVETRAIN

Type	ohv inline 4
Bore x stroke, mm	73.7 x 87.4
Displacement, cc/cu in.	1493/91.0
Compression ratio	9.0:1
Bhp @ rpm, net	55.5 @ 5000
Torque @ rpm, lb-ft	73 @ 2500
Fuel requirement	premium, 96-oct
Transmission	4-sp manual
Gear ratios: 4th (1.00)	3.90:1
3rd (1.43)	5.58:1
2nd (2.11)	8.23:1
1st (3.41)	13.30:1
Final drive ratio	3.90:1

CALCULATED DATA

Lb/bhp (test weight)	39.1
Mph/1000 rpm (4th gear)	16.2
Engine revs/mi (60 mph)	3700
R&T steering index	0.70
Brake swept area, sq in./ton	214

ROAD TEST RESULTS

ACCELERATION

Time to distance, sec:	
0–100 ft	3.8
0–500 ft	10.8
0–1320 ft (¼ mi)	20.1
Speed at end of ¼ mi, mph	67.0
Time to speed, sec:	
0–30 mph	4.8
0–50 mph	10.7
0–60 mph	15.5
0–80 mph	36.5

SPEEDS IN GEARS

4th gear (5100 rpm)	83
3rd (6000)	70
2nd (6000)	48
1st (6000)	29

FUEL ECONOMY

Normal driving, mpg	27.5

BRAKES

Minimum stopping distances, ft:	
From 60 mph	217
From 80 mph	326
Control in panic stop	fair
Pedal effort for 0.5 g stop, lb	50
Fade: percent increase in pedal effort to maintain 0.5g deceleration in 6 stops from 60 mph	see text
Overall brake rating	poor

HANDLING

Speed on 100-ft radius, mph	33.2
Lateral acceleration, g	0.737
Speed thru 700-ft slalom, mph	50.1

INTERIOR NOISE

All noise readings in dBA:	
Constant 30 mph	72
50 mph	81
70 mph	86

SPEEDOMETER ERROR

30 mph indicated is actually	31.0
60 mph	61.0
70 mph	71.0

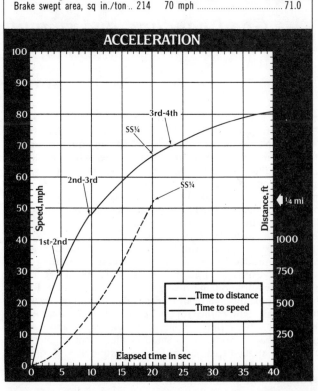

of very long stopping distances.

Keeping in mind the petite dimensions of the car, it's no surprise that it is not comfortable for anyone over about 6 ft. The quality of the interior finish is high and the soft top fits better than the last time we tested a Midget. There is still plenty of noise, mechanical and wind, and that means the car seems to be going fast at almost any speed. With the top down, and this is still not a simple task although it has been improved, the speed sensation goes up dramatically. And perhaps this is what the car is all about. All the elements are there for the young enthusiast: wind in the face, lots of gear changing, and a plethora of sports car noises. Power becomes rather secondary in this type of driving and responsiveness is of primary importance. The Midget also meets many of the other rudimentary requirements of a sports car: it has a proud name earned in road racing competition, there is an awareness of things mechanical going on all about you from the note of the exhaust to the whine of the gearbox, and there are gauges for monitoring vital functions rather than warning lights. All of these qualities come together in making a sports car something that demands to be driven with verve and expertise.

All in all, despite its outdated design and time-worn engineering, the Midget still provides that most important quality: it's fun to drive in the sports car tradition.

AUSTIN-HEALEY SPRITE/MG MIDGET, 1958-1970

Good handling and simplified design characterize these inexpensive sports cars

BY THOS L. BRYANT

FORMER R&T PUBLISHER John Bond once wrote, "To be successfully manufactured and sold, for minimum price, a car has to be of utmost simplicity in order to keep manufacturing costs low. Sports cars will never sell in the same quantity as family-type cars, so the argument that a sports car can sell as cheaply as a VW is not entirely valid" ("Miscellaneous Ramblings," July 1958). John was discussing the introduction of the new Austin-Healey Sprite, announced in May 1958 and destined to be as successful in its own way as the medium-size Austin-Healey 100 and 3000.

In the flush of success with the Big Healeys, Donald Healey decided to produce a smaller, more economical sporting car in the least expensive price range. As with the Big Healey, he based the Sprite on an existing sedan, the Austin A35, from which he took the 948-cc overhead valve 4-cylinder engine. This unit was slightly tuned for the Sprite with twin SU carburetors, special valve springs, improved exhaust valves and modified crankshaft bearings. This small displacement engine put out 48 bhp at 5000 rpm and 52 lb-ft of torque at 3300 rpm.

A 6-in. single dry-plate clutch connected the engine to a 4-speed manual gearbox with synchromesh on 2nd, 3rd and 4th. Ratios were 3.63 for 1st, 2.37 for 2nd, 1.41 for 3rd and 1.00:1 in 4th. The rear axle ratio was 4.22:1.

The rack-and-pinion steering system went from lock to lock in just 2.3 turns, giving the car very quick steering that took some getting used to initially. In our first road test of a Sprite (August 1958) we said, "In fact, the steering is nearly perfect for the purpose, and light and accurate besides."

The front suspension was also taken from the Austin A35 and consisted of lower A-arms and coil springs with lever-type hydraulic damper arms providing the upper suspension link. The solid rear axle was located by quarter-elliptic springs and lever-type shock absorbers. The Sprite came with 7-in. diameter drum brakes and 13 x 3½-in. steel wheels shod with 5.20-13 bias-ply tires at all four corners. This all worked quite well and cornering characteristics were close to neutral with a small amount of understeer at moderate speeds. It was very difficult to get the rear end to break loose and when it did it could be controlled quite easily.

While all of the drivetrain components and underpinnings were very conventional, the Sprite body was uniquely odd in appearance—so much so, in fact, that after awhile it began to look rather attractive! This apparent contradiction struck most of us in those early Sprite days as most of the motoring journals of the time had unkind remarks for the shape and design of the car. Soon, the bulging headlights and smiling grille were characterized as giving the Sprite a frog-like appearance and it became known as the frog-eye or Bugeye Sprite. Originally, Donald Healey's design had called for retractable headlights but when BMC actually began producing the car, this design was deemed too expensive and they were left perched in the middle of the hood.

The body of the car was a pressed steel shell and the entire or perhaps tertiary importance.

The Sprite carved a new niche for itself in the sports car world, offering superior and economical performance for a price comparable to the least expensive sedans of the day. Almost immediately, Sprites began to appear at sports car races all over the world and soon dominated H production racing in the U.S. Who could ever forget those days of the late Fifties and early Sixties when a gaggle of the little cars would take off sounding for all the world like a horde of angry bees chasing a hive-robbing bear! Not only did club racers turn to the Sprite; such big names as Stirling Moss, Walt Hangsen, Bruce McLaren,

Austin-Healey Sebring Sprite.

Sprite dashboards weren't known for their over-instrumentation.

front end was hinged at the cowl to give easy access to the engine, steering and front suspension. The rest of the Sprite was fairly neat and clean with a simple, slightly sloping rear deck unmarred by a trunk lid. Entry to the trunk was from within the car and although the seatbacks were hinged to move forward, access to the luggage space was limited. John Bond described the appearance of the Sprite as a "hybrid of TR-3, Berkeley and Crosley Hotshot" and we are inclined to let it go at that.

The original Sprite was a classic roadster with removable soft top and side curtains. It was reasonably roomy inside with more pedal and leg room than the MGA but getting inside to make use of that room could be rather challenging, especially for a driver more than 5 ft 10 in. tall. The top came down far enough to restrict vision to the sides somewhat and rainy weather often pointed up a few places where the fit was not absolutely water tight, but then this was a sports car in the traditional sense of the word and comfort was of secondary

Briggs Cunningham and others competed in works cars or the Sebring Sprites of John Sprinzel.

Sprite Mark II

THREE YEARS after the introduction of the original Sprite and some 49,000 cars later, the Mark II version was unveiled in May 1961. Gone were the frog eyes, replaced by conven-

Austin-Healey Sprite Mark II.

tionally located headlights in fixed fenders, along with a wider grille and a hood that opened without lifting the entire front body just like most other cars. The rear was also restyled with a trunk lid that opened from outside and a 12-in. cut in the back of the cockpit for luggage or a small child.

There were also changes to the engine, including an increase in the compression ratio from 8.3 to 9.0:1, larger throats in the SU carburetors, larger intake valves and a change in the exhaust valve timing. The result was more power (50 vs 48 bhp) without loss of torque. The gearbox was also revised; the close-ratio gears used on the Sebring Sprites in competition and available as an option for a year or so were now standard in the new car.

To the avid Bugeye Sprite fanatic, the Mark II was an abomination. To the rest of the motoring enthusiast world, however, the new car was a definite improvement: "Of course, the new model may be accused of some minor loss of personality, but no one can deny that the Sprite II is better looking" (R&T, August 1961). The Mark II was also a more convenient car to live with because of its outside opening trunk and better use of space. R&T concluded that the Mark II, like the original Sprite, "offers more fun per dollar than anything we have driven for a long time."

Only a month after the announcement of the Sprite II, the same car was brought out under the MG banner as the Midget. The only differences were in the nameplates and a few minor trim details and this practice was to continue until the eventual demise of the Sprite name in 1970, when the Midget continued on by itself.

Sprite 1100

FOUR-AND-a-half years after the introduction of the Sprite, a new 1100 model was shown at the London Motor Show in October 1962. The new Sprite (and Midget) had been upgraded with an increase in engine displacement from 948 to 1098 cc, which raised the horsepower rating from 50 to 55 at 5500 rpm. The clutch diameter was increased by 1 in., front brakes were now disc rather than drum, and reshaped and more thickly padded seats plus the addition of carpeting helped to make the Sprite more comfortable.

The additional engine displacement was the result of an increase in bore (2.54 vs 2.48 in.) and a new, longer-stroke crankshaft very similar to the one used in the MG 1100 sedans. The displacement increase made the Sprite more pleasant to drive because of the extra torque available (61 lb-ft at 2500 rpm compared with 52.5 at 2750 in the Mark II).

Although the new seats were certainly more comfortable and the noise level had been reduced by the carpeting, the Sprite was still characterized by many of its original features: no wind-up windows, separate key and starter (the latter a pull cable rather than button), no door locks or outside door handles and no glovebox. The only lockable portion of the entire car was the trunk.

In terms of performance, the Sprite 1100 was a much better car. Top speed improved by only a few mph but the change in the torque characteristics offered top gear hill climbing and easier passing than in previous Sprites. This new found performance could be used with increased confidence as a result of the change to front disc brakes. The former drum brakes were marginal when the car was being driven near the limit. Handling was unaffected by any of these changes and the Sprite (and Midget) was still one of the most responsive cars on the road.

Perhaps more amazing than anything else was the fact that the new Sprite buyer was still getting all of this for less than $2000! In the period from May 1958 to October 1962, the Sprite's list price had only gone from $1795 to $1985. And *Road & Track* was still saying that this car offered more fun per dollar than any other, even after 4½ years.

Sprite Mark III

MARCH 1964 brought the introduction of the Mark III Sprite (Mark II Midget which was one number behind) and

this new model reflected the need to bring the Sprite up to date with the modern sports car. Major improvements included wind-up windows, swiveling vent wings, updated instrument panel and further improvements in interior trim.

The rear suspension came in for revision with a change from the quarter-elliptic springs to semi-elliptics to reduce the tricky roll oversteer inherent in the original design. There was also a minor improvement in engine performance (stepped up to 59 bhp) through improved manifolding and the use of the MG 1100 cylinder head.

Sprite Mark IV/Midget Mark III

ONCE AGAIN the London Motor Show was the arena for the display of the new Sprite/Midget, this time in 1966. The car now became a thoroughly modern sports car with a proper

convertible top that could be raised and lowered without dismantling. Another engine transplant had taken place and the Sprite/Midget now had a 1275-cc powerplant similar to that used in the Mini Cooper S but detuned from 75 bhp to 65. This detuning allowed lower production costs because a normal forged crankshaft could be used in place of the more expensive nitrided steel crank of the S for example, while still maintaining the Sprite's reputation for reliability and long life. The net increase of 6 bhp over the Mark III Sprite (Mark II Midget) was enough to make a surprising improvement in performance as the new Sprite/Midget would accelerate to 60 mph in 14.7 seconds versus 18.3 sec for the previous model.

The crisp handling which had always been a characteristic

of Sprites remained and there was still some roll oversteer built into the rear suspension that made the car great fun to drive sideways. The ride was still a bit jouncy over uneven surfaces but it was truly sporting in nature and aficionados of the breed didn't mind it at all and in fact felt (and still feel!) that it was necessary or you might as well have been driving a large sedan.

The Sprite/Midget cars remained little changed from this setup until 1975. Of course, the Sprite version was discontinued in 1970. The Midget continued without any drastic changes except for the addition of more and more emission controls and safety items such as the over-large bumpers of today, until the 1975 model which received the 1493-cc Triumph Spitfire engine with slightly different exhaust manifolding. Emission controls had become stringent enough that the increased displacement was capable of putting out 55.5 bhp at 5000 rpm.

For the purposes of our discussion here, we have elected to cut off our report with 1970 for two reasons: that was the last year for both the Sprite and the Midget; and the increasing emission and safety regulations have taken some of the fun out of driving the more modern Midgets. They still have reasonable handling, are relatively lightweight and thrifty of fuel, but in talking about a Used Car Classic we also have to give consideration to purchase price and post-1970 car prices are generally out of the bargain class.

Car selection tips

As with most British sports cars of the time, the early Sprite/ Midget is a remarkably sturdy and simple car. The engine and transmission having come from a sedan of some years' standing, they were time-tested and of proven reliability. The twin SU carburetors have a tendency to be touchy in adjustment so many times a car that does not seem to be working properly may just need a delicate hand applied to the carburetors. Also, the linkage runs right into the carburetor throat and there have been cases where the opening becomes worn, allowing air to seep in and upset the mixture. The use of some rubber grommets can cure this malady. One of the few weak spots in the engine is the center main bearing on the crankshaft. If it's at all suspect, replace it.

All of the Sprite/Midgets covered in our time period have a non-synchromesh 1st gear. While the gearbox is sturdy, a heavy-handed driver can wreak damage on the unit, especially 1st, and it may be necessary to consider rebuilding it. On the other hand, we have been told by Sprite owners that they have put considerably more than 100,000 miles on their cars without the gearbox showing signs of wear.

Rust is one of the most important things to look for in buying any car and it's true of the Sprite too. The potential buyer may be fortunate and find a car that has a number of holes drilled in the rocker panels and the bottom edge of the hood (Bugeye models)—these are to permit water to drain out and prevent rusting.

Prices for used models vary from area to area of the country. The Bugeye is rapidly approaching classic status and the prices are beginning to reflect this. However, a running Bugeye that needs restoration can occasionally be found in the $600–800 range in southern California, but the usual price is $200–400 more than that. The fully restored Bugeye can bring as much as $2000 and more. Prices will generally be somewhat higher

TYPICAL ASKING PRICES

Year & Type	Price Range
1958–61 Austin-Healey Sprite	$600–1200
1961–62 Austin-Healey Sprite II & MG Midget	$500–1200
1963 Austin-Healey Sprite 1100 & MG Midget 1100	$400–1200
1964–65 Austin-Healey Sprite III & MG Midget II	$500–1500
1966–70 Austin-Healey Sprite IV & MG Midget III	$800–1800

PERFORMANCE DATA
From Contemporary Tests

	1958 Austin-Healey Sprite	1961 Austin-Healey Sprite II	1963 Austin-Healey Sprite 1100	1967 MG Midget III
0–60 mph, sec	20.8	19.6	18.3	14.7
0–80 mph, sec	35.5*	49.0	42.5	31.0
Standing ¼ mi, sec	21.8	21.5	20.9	19.9
Avg fuel economy, mpg	34.0	34.0	33.0	24.0
Road test date	8-58	8-61	8-63	9-67

*0–70 mph

BRIEF SPECIFICATIONS

	1958 Austin-Healey Sprite	1961 Austin-Healey Sprite II	1963 Austin-Healey Sprite 1100	1967 MG Midget III
Curb weight, lb	1460	1540	1560	1560
Wheelbase, in.	80.0	80.0	80.0	80.0
Track, f/r	45.4	45.8	47.2	46.3
	44.8	44.8	45.0	44.8
Length	137.0	136.0	138.0	137.4
Width	54.0	54.0	54.0	56.5
Height	48.0	48.8	47.8	48.6
Fuel capacity, gal.	6.0	7.2	7.2	7.5
Engine type	ohv inline 4	ohv inline 4	ohv inline 4	ohv inline 4
Bore x stroke, mm	63.0 x 76.2	63.0 x 76.2	64.5 x 83.8	70.6 x 81.3
Displacement, cc	948	948	1098	1275
Compression ratio	8.3:1	9.0:1	8.9:1	8.8:1
Bhp @ rpm, SAE gross	48 @ 5000	50 @ 5500	55 @ 5500	65 @ 6000
Torque @ rpm	52 @ 3300	52.5 @ 2750	61 @ 2500	72 @ 3000
Gearbox	all models: 4-speed, non-synchro 1st gear			
Final drive ratio	4.22:1	4.22:1	4.22:1	4.22:1

in other parts of the country where these particular models are less readily available. Prices on newer Sprites and the MG Midgets will run anywhere from $800 on up, peaking at about $1800.

The Sprite/Midget series of cars is perhaps one of the most logical ones for inclusion in our Used Car Classic series of reports. They are simple, relatively easy to maintain, and, just as when they were new, return more driving fun per dollar than just about any sports car we can think of. If you are the proper size to fit inside one, you couldn't do much better. 🞅

COMPARISON R&T ROAD TEST

THE AFFORDABLES

A head-to-tail confrontation between six moderately priced sports cars: Fiat 124 Spider, Fiat X1/9, MGB, MG Midget, Triumph Spitfire and Triumph TR7

PHOTOS BY JOE RUSZ & JOHN LAMM

NOT TOO MANY years ago you could have bought a lot of sports car for $3500. The Datsun 240Z, Fiat 124 Spider, Porsche 914, MGB and Triumph TR6 all hovered around that price. And if you only had $2500 or so, you could buy more basic sports cars like the Triumph Spitfire and the MG Midget. If you've checked new car prices recently, however, you have to be astonished and dismayed at the havoc world inflation has wrought. The least expensive sports car sold in America in 1976, the MG Midget, lists for $3949, and over the past four years the 280Z and 914 have almost doubled in price!

The popularity of our Used Car Classic series and the record or near-record sales posted by British Leyland and Fiat in 1975 attest to the fact that a lot of car enthusiasts are also sports car enthusiasts. But more so today than ever before they are carefully looking at value for the dollar. Which is where R&T comes in. With what has happened to prices the last few years, coupled with the introduction of completely new sports cars like Fiat's

X1/9 and Triumph's TR7 we decided it was time to take a fresh look at the various options available to a potential sports car buyer. Obviously this runs the gamut from the Porsche Turbo, the Maserati Merak and similar cars only the very rich can afford, all the way down to the diminutive Midget. What we were most interested in, however, were those cars within the reach of the average R&T reader.

After much discussion we decided to limit the test to those sports cars costing less than $6000. If that seems high (it did to us at first) consider this: only seven sports cars fall below that figure and they are Fiat 124 Spider, Fiat X1/9, MGB, MG Midget, Triumph Spitfire, Triumph TR6 and Triumph TR7. We included six of those seven cars in our comparison test, leaving out the TR6 because it is not sold in California and we ran into insurmountable logistical problems in trying to borrow one.

The price spread of the six cars—from $3949 for the Midget to $5845 for the Fiat 124 Spider—is larger than usual for our

comparison tests. But the cars are quite dissimilar with features and characteristics that might strongly appeal to some drivers but not to others. And even among the staff there wasn't universal agreement as to whether or not a higher price tag would necessarily mean a better car.

Three of the cars—MGB and Midget and Spitfire are what we would consider classic roadsters in the Best British Tradition. The 124 would probably be considered by most to be a more civilized sports car done up with the usual Italian flair.

The TR7 represents a complete break from the past. Though based on an existing sedan series—as sports cars usually are—it is a new car for all practical purposes that draws on contemporary engineering practice to combine its performance and handling with a high degree of comfort. It's the only car in the group without a removable top. But British Leyland will remedy that situation sometime soon by offering a large sunroof similar to the Renault 17 Gordini coupe/convertible design.

The Fiat X1/9, the only mid-engine car in this group, is also a trendsetter in some respects. It's the first good-looking open mid-engine roadster available to buyers of moderate means.

The six cars differ greatly in specifications as well as vintage. Both MGs and the Spitfire have undersquare overhead-valve 4-cyl engines; the TR7 and the X1/9 have decidedly oversquare 4-cyl engines with a single overhead camshaft; the 124's 4-cyl is an efficient dohc design. In chassis design the 124, both MGs and the TR7 have unit bodies with live rear axles, the Spitfire has a separate body and frame with independent suspension and the

X1/9 is unique in having a midship engine, naturally with independent suspension. To transmit power the 124 has a 5-speed gearbox while all the others have 4-speeds. Overdrive is an option with the MGB and the Spitfire, however, and for 1977 Triumph's TR7 will be available with automatic. When that happens the TR7 will replace the 280Z as the lowest priced sports car offered with that option. Both Fiats have 4-wheel-disc brakes; all the English cars have front discs and rear drums. Probably the greatest similarity between the six cars is their common purpose of providing driving pleasure.

None of the six cars is a new model, all having been tested by us at least once before in separate road tests. Although their basic character has not changed there have been detail changes in all of them since we last tested them. Still, there's nothing like taking a trip and comparing them head to head for making an accurate assessment of the practical, functional and aesthetic qualities of each and to arrive at an overall rating of preference.

So a test crew of eight (including Editor-at-Large Henry Manney whose sometimes caustic wit and overall irreverence tempered the more routine approach of the others) took turns behind the steering wheels and in the passenger seats during a 2-day jaunt that put the cars through the sort of use their owners might. We filled the fuel tanks at San Juan Capistrano and headed out the Ortega Highway to Lake Elsinore. This highway is a nicely surfaced, 2-lane route with a delightful variety of turns and hills as it winds through some of southern California's finest country. From Lake Elsinore we headed southeast through Temecula and Warner Springs into the Anza-Borrego Desert where we spent the night. Next day we headed back toward San Juan Capistrano over the desert and along some wonderfully twisty roads at speeds dictated by road conditions and car capability rather than artificial limits, passing through Banner, Rincon Springs and Pala along the way. At Oceanside we hooked up with the San Diego Freeway for a leisurely 55-mph drive back to San Juan where we again topped up the fuel tanks. A large portion of this trip did involve sports car country and we drove these cars hard for miles on end so the trip fuel economy figures are a bit on the low side. Prospective buyers of any of the six can expect to get more miles per gallon in daily use.

Back at the office we set about scoring the cars. Each driver had a score sheet with 20 different categories in which to rate the cars, and after the trip each sat down immediately with clipboard in hand, the six cars nearby and the pages of notes each had made at our various meal, rest and driver-change stops on the trip. All the categories could be scored on a 1 to 10 basis, 10 being the score of a topnotch performance and 1 being the lowest score possible. These scores were then totaled for each driver and for the entire group to get an overall rating score for each car.

In addition, each driver was asked to rank the cars in the order of his or her personal preference—disregarding, if necessary, the separate ratings of the car's various aspects. Here's how the ratings turned out. Of a possible 1600 points, the scoring was as follows: Fiat X1/9 1371; TR7 1247; Fiat 124 1214; MGB 940; Spitfire 920; Midget 761. In individual driver scoring the X1/9 garnered the highest number of points from all eight drivers and everyone rated it his personal favorite. Next came the TR7 with six drivers rating it second in points and two drivers placing it third. For personal perference the TR7 scored six second places, one third and one fourth. The 124 was a solid third-place finisher. Six drivers had it third in points, two drivers had it second. These same two drivers also gave it second place in personal preference while one of the other six had it fifth on his most favorite list.

A clear step below the two Fiats and the TR7 came the MGB and the Spitfire. In the points standings both had four fourth-place finishes and three fifth places. One driver had the MGB in a third-place tie with the 124 and another had the Spitfire last. On the personal favorites list four drivers had the MGB fourth and four had it fifth. The Spitfire was well enough liked to be scored third favorite by two drivers; the other six voters were evenly split between fourth and fifth places.

Finally we come to the Midget, clearly trailing the pack by a wide margin. Seven of the drivers had it last in points; its lone fifth place was by one point over the Spitfire. And all eight drivers had it last in their personal ratings.

Fiat X1/9

NOW LET's take a closer look at all six cars. The Fiat X1/9 was rated best in 13 of the 20 categories: handling, ride, quietness, braking, steering, gearbox, driving position, exterior and interior styling, exterior and interior finish, luggage space and loading and solidarity of its body structure. Its lowest placing in any category was a third in instrumentation and even there it scored a total of only three points less than the top-rated TR7. Overall balance is definitely the X1/9's forte and it's obvious that Fiat engineers really did their homework in coming up with what we consider the most significant sports car since the Datsun 240Z.

The X1/9 has the smallest engine but it makes every cubic centimeter count as only the 124 and TR7 out performed it to any degree. And what the X1/9 lacks in all-out straight-line acceleration it more than makes up for with its impressively smooth, quiet

and easy revving engine.

The ride, steering, braking, handling and roadholding are all first rate with the usual Fiat expertise applied to the always potentially outstanding mid-engine layout. There's an almost perfect blending of springing, damping and wheel travel, so the ride, though firm, is not harsh like the 914's. The body is solid and rattle free and the steady-state cornering is on a par with much more expensive sports cars. But what makes the X1/9 such an excellent road car are its transient characteristics. There's minimal roll and even if you lift off the throttle in a tight turn there's no abrupt change in attitude as sometimes happens with rear-heavy cars. The front tucks in a bit but there's no abrupt change to oversteer.

Designed by Bertone, the X1/9 has clean, taut lines and an aggressively agile appearance marred only by ugly safety bump-

ers. The handsome interior is comfortable for driver and passenger but we are a bit dismayed that Fiat has fitted only lap belts and not a safer lap-shoulder belt system. One other complaint: Engine accessibility seems rather appalling and in these days of $22 an hour labor rates in sections of the country this can be important.

The X1/9 is a thoroughly professional effort: small, light, frugal, immensely entertaining to drive and yet comfortable and practical. It's no wonder that driving the X1/9 elicited such comments as, "Bloody nice little car," and, "It sounds like it's having fun too."

Triumph TR7

THE TR7 is proof that in between labor strikes British Leyland can design modern sports cars when it wants to. It placed highest in five categories: engine, controls, instrumentation, ventilation and ease of top removal and installation. The latter was not voted facetiously despite what you might think. The car we tested didn't have a sunroof but we studied a photo of the sunroof car and a description of its operation found in the owner's manual before voting the TR7 best in this category. In every other category except outward vision the Triumph was a consistent second or third place finisher. In outward vision the Triumph fell to last place because the forward extremities are largely invisible and the wide sail panels interfere with rear-quarter vision. We were somewhat surprised after totaling the sums to discover the TR7 rated third in exterior styling, but it scored fewer points in this category than in any other area under consideration except outward vision and it was a distant third behind the two Fiats. The Triumph's strange and stubby appearance sort of grows on you after a while but one of the staff wags summed it up best when he said the TR7 looks like an X1/9 with a bad case of the mumps.

TR7 answers the cry for an open car with this sliding sunroof.

Triumph Spitfire Mk IV 1500.

MG Midget Mark III.

If the exterior styling is a bit offputting, Triumph certainly did the right things inside. The interior is spacious and luxurious, the gauges and controls are all in the right places and the fabric-covered seats are comfortable with easily adjustable backrests. Our 1975 road test car had foot pedals located so enthusiastic drivers could heel and toe easily. In this latest example, however, the space between the brake and throttle pedals has been widened, making double-clutching while downshifting a difficult proposition. Strange.

The TR7's engine, at least in the 2-carburetor 49-state version we tested, is by far the strongest runner in this group. The Triumph isn't particularly quick off the mark because 1st gear is surprisingly tall and this requires a bit of clutch slip to get the car moving from rest. But once underway the TR7 quickly puts a wide expanse of pavement between itself and the next most lively car, the 124 Spider. The gear change suffers from a recalcitrant reverse gate and a slight graunch whenever you downshift into 1st gear but is otherwise quite satisfactory. The brakes are very good both for pedal modulation and stopping distances and the handling is very driver friendly: understeer is the prevailing trait but just enough to keep the novice driver safe and not enough to bother the expert. The steering is neither terribly quick nor light but it is precise and it does lighten up at speed. The ride is well controlled and comfortably supple but at high speeds on undulating pavement the front end gets a little floaty.

Even though it blazes no new trails in the evolution of the sports car, we think Triumph has done a generally nice job with the TR7. The handling-ride-braking combination is most pleasing, it's reasonably light and economical of fuel, it's an exceptionally comfortable car with a wide, deep and regularly shaped trunk, and it's entertaining to drive. You couldn't ask for much more than that.

Fiat 124 Spider

ALTHOUGH IT dates back to 1968 the Fiat 124 Spider is still a very civilized sports car and of the four "vintage" cars in this test it has aged the most gracefully. That was made very clear in the scoring because the 124 was a close third behind the considerably newer TR7. The 124 was voted best in two categories—heating and outward vision—and scored no lower than third

Fiat 124 Sport Spider.

in any area except for fourth place finishes in ventilation and braking (more about that later) and a fifth in driving position.

Keeping up with emission requirements has severely weakened Fiat's once strong dohc 4-cyl, but nevertheless it is still a satisfying piece of machinery: quiet, very smooth for a 4-cyl and willing to rev to its 6500-rpm redline. And in this group of cars it was still second fastest. The 124 has the only 5-speed gearbox; although it requires longer throws than the British 4-speeds, the shifter is quite precise in its action.

Probably the most controversial aspect of the 124 is the driving position. Either you love driving with your arms extended 'way out and the pedals close or you hate it. In this case it was a split decision, four voting yea and four nay. Otherwise there's little to complain about inside. The instruments are easy to read, the controls well laid out and stowing and raising the soft top is utter simplicity. And the top provides far better weather protection than any of the British roadsters.

The Fiat's basic understeer and low-speed steering heaviness are not appreciated at first; after miles of fast driving the Fiat's excellent transient cornering and good absolute adhesion make it a most enjoyable car. It's a supple riding car in the best Italian tradition with exceptional axle control over rough pavement.

The brakes are an enigma. They achieved creditable stopping distances in our simulated panic stops, although the front discs locked easily and unequally causing the car to pull to the left. They faded more than we expected for an all-disc design but not to an alarming degree. But while driving through Anza-Borrego our first day out, the brakes faded completely away during one hard downhill stop from 70 mph and were never completely right thereafter. This is obviously not characteristic behavior because we have found Fiats to have better than average brakes.

There's no disputing the fact that the 8-year-old 124 is outdated in some ways. However, the Italians design sports cars with so much feeling for the car and the driver that many of their cars are endowed with a timeless quality that makes them enjoyable to drive despite their age.

Triumph Spitfire

IN THE individual points scoring the Spitfire's best showing was four fourth-place finishes. On the personal preference lists, however, the Spitfire found itself with two third place votes, one driver placing it ahead of the TR7 and one ahead of the Fiat 124. And in the overall scoring it had only 20 fewer points than the MGB. The Spitfire's highest placing in any of the individual categories was a second place in braking. It also scored four worsts: instrumentation, ventilation, ease of top removal and installation and body structure, but only in the latter was it far

	Fiat 124 Spider	Fiat X1/9	MG B	MG Midget	Triumph Spitfire	Triumph TR7
GENERAL DATA						
Basic price	$5845[1]	$4947	$4795	$3949	$4295	$5649
Price as tested[2]	$6045	$5147	$4995	$4299	$4495	$5847
Layout	front engine/ rear drive	mid engine/ rear drive	front engine/ rear drive	front engine/ rear drive	front engine/ rear drive	front engine/ rear drive
Curb weight, lb	2255	2045	2275	1775	1750	2220
Test weight, lb	2635	2420	2645	2170	2120	2580
Weight distribution (with driver), f/r, %	51/49	41/59	50/50	52/48	53/47	55/45
Wheelbase, in.	89.7	86.7	91.1	80.0	83.0	85.0
Track, f/r	53.2/52.0	52.5/52.9	49.0/49.3	46.3/44.8	49.0/50.0	55.5/55.3
Length	163.1	158.5	158.3	141.0	156.3	164.5
Width	63.5	61.8	59.9	54.0	58.5	66.2
Height	49.2	46.1	50.9	48.3	45.6	49.9
Wheel size	13x5	13x4½	14x4½	13x4½	13x4½	13x5½
Tires	Goodyear G800 Grand Prix 165HR-13	Michelin XAS, 145HR-13	Dunlop SP68, 165SR-14	Pirelli Cinturato CF67, 145SR-13	Goodyear Custom G800 Rib, 155SR-13	Michelin X, 175/70HR-13
Fuel capacity, U.S. gal.	11.4	12.7	14.0	7.5	8.8	14.5

[1] Price varies according to port of entry.
[2] As-tested price includes: for the Fiat 124 Spider, AM/FM radio ($200); for the Fiat X1/9, AM/FM radio ($200); for the MGB, AM/FM radio ($150), tonneau cover ($50); for the MG Midget, AM/FM radio ($150), wire wheels ($135), tonneau cover ($50), anti-roll bar ($15); for the Triumph Spitfire, AM/FM radio ($150), tonneau cover ($50); for the Triumph TR7, AM/FM/tape ($198).

Raising or lowering the top on the English roadsters was an exercise in frustration.

behind the fifth place car.

The Spitfire surprised a lot of people. Drivers commented, "The handling is damned impressive," "The suspension, steering, ride and roadholding are much better than I suspected," and "It's a surprisingly good car." On external styling one driver said it best: "The Spitfire has aged very well and has grown into the safety age (it was introduced in the U.S. in 1963) better than any of the others that were around before The Great Silliness. And when you consider what the first Spitfires looked like this is a great improvement."

The Spitfire is the only car with a separate body and frame and is not as solid feeling as the unit-body cars. The hood shakes, the chassis seems to flex a bit and there is excessive steering feedback on rough roads through the otherwise excellent rack-and-pinion steering. The Spitfire's independent rear suspension is a relatively unsophisticated swing-axle design but some clever re-engineering in 1971 improved the handling enormously. Only the main spring leaf of the Spitfire's transverse leaf spring is clamped to the differential so that it can contribute roll stiffness; the other leaves contribute practically nothing to roll stiffness. Furthermore, the whole spring assembly has a higher 2-wheel jounce rate—roughly double what it was—so that camber change under acceleration and braking is reduced. This design decreases the oversteer and jacking of the rear wheels that was characteristic of earlier Spitfires.

The Spitfire's engine is rough as old boots when revved above 5000 rpm (redline is at 6000) and it's noisy as well in the upper speed ranges, but it provides peppy performance otherwise, especially in the lower three gears.

The Spitfire's seats drew mixed reviews as the seat cushions are short and don't provide enough thigh support. It also has less instrumentation than the other cars but otherwise the interior is

PERFORMANCE

	Fiat 124 Spider	Fiat X1/9	MG B	MG Midget	Triumph Spitfire	Triumph TR7
Lb/bhp (test weight)	30.6	36.9	42.3	39.1	36.9	28.7
Acceleration: time to speed, sec						
0–30 mph	4.9	5.3	5.5	4.8	4.8	4.3
0–60 mph	14.8	16.3	18.3	15.5	15.3	11.5
0–70 mph	20.5	23.7	26.5	22.8	22.3	15.7
0–80 mph	30.4	34.7	39.0	36.5	32.3	20.9
Standing ¼ mile, sec	20.0	20.4	21.5	20.1	20.2	18.5
Speed @ ¼ mile, mph	69.5	66.5	64.5	67.0	67.5	76.0
Top speed, mph	100	90	90	83	94	105
Trip fuel economy, mpg	22.0	26.0	19.5	29.0	25.0	22.5
Braking: stopping distance, ft, from						
60 mph	170	151	177	189	167	169
80 mph	285	280	320	321	304	289
Control in panic stop	fair	excellent	fair	very good	very good	excellent
Pedal effort for 0.5g stop, lb	25	38	25	38	50	32
Fade, % increase in pedal effort in six 0.5g stops from 60 mph	40	58	60	97	10	9
Overall brake rating	good	very good	fair	fair	good	very good
Cornering capability, g	0.737	0.791	0.698	0.737	0.731	0.760
Speed thru 700-ft slalom, mph	55.8	57.4	53.0	50.1	58.1	58.6

INTERIOR NOISE

	Fiat 124 Spider	Fiat X1/9	MG B	MG Midget	Triumph Spitfire	Triumph TR7
All noise readings with top up, dBA:						
Idle in neutral	59	59	61	61	61	60
Maximum, 1st gear	87	82	87	87	87	88
Constant 30 mph	72	67	70	72	74	67
50 mph	75	72	75	81	79	71
70 mph	83	79	82	86	87	77

ENGINE & DRIVETRAIN

	Fiat 124 Spider	Fiat X1/9	MG B	MG Midget	Triumph Spitfire	Triumph TR7
Engine type	dohc inline 4	sohc inline 4	ohv inline 4	ohv inline 4	ohv inline 4	sohc inline 4
Bore x stroke, mm	84.0 x 79.2	88.0 x 55.5	80.3 x 89.0	73.7 x 87.4	73.7 x 87.4	90.3 x 78.0
Displacement, cc	1756	1290	1798	1493	1493	1998
Compression ratio	8.0:1	8.5:1	8.0:1	9.0:1	9.0:1	8.0:1
Bhp @ rpm, SAE net	86 @ 6200	65.5 @ 6000	62.5 @ 5000	55.5 @ 5000	57.5 @ 5000	90 @ 5000
Torque @ rpm, lb-ft	90 @ 2800	67 @ 4000	72 @ 2500	73 @ 2500	75 @ 2500	106 @ 3000
Carburetion	one Weber (2V)	one Weber (2V)	one Zenith-Stromberg (1V)	one Zenith-Stromberg (1V)	one Zenith-Stromberg (1V)	two Zenith-Stromberg (1V)
Fuel requirement	unleaded, 91-oct	unleaded, 91-oct	unleaded, 91-oct	premium, 96-oct	premium, 96-oct	regular, 91-oct
Transmission	5-sp manual	4-sp manual	4-sp manual	4-sp manual	4-sp manual	4-sp manual
Final drive ratio	4.30:1	4.08:1	3.90:1	3.90:1	3.89:1	3.63:1
Engine speed @ 60 mph, rpm	3330	4100	3320	3700	3500	3300

much more roomy and livable than the Spitfire's direct competitor, the Midget. Overall we'd have to say that although it has its failings, these days the Spitfire is the cheapest way to get into an acceptable new sports car.

MGB

WE'VE HEARD that British Leyland is simply letting the MGB run its historic course; when it can't be sold anymore they'll drop it and that's that. The car seems to bear it out. Meeting the U.S. crash-safety regulations was done by laying an ugly, add-on instrument panel over the existing one and the little bit of styling facelift has been done in a haphazard way.

It's truly a car of the past. Everywhere there's evidence of a sports car designed and built in the traditional manner—in a rather homemade way, to be blunt...

Think you've heard that said somewhere before? You're right. The previous two paragraphs are a direct quote from a 1971 test comparing the Datsun 240Z, Fiat 124 Sport, Opel GT, MGB GT and Triumph GT6. And it's truer today than it was five years ago. To meet bumper crash regulations British Leyland grafted on ugly rubber appendages to both ends of the car. These look bad enough but because the MGB still couldn't meet bumper height requirements the body was raised on its springs about 3 in. adding injury to insult. Not only does the MGB look like a candidate for the Baja 1000 but the car rolls excessively and the handling has deteriorated severely.

If the MG's pushrod engine never produced a lot of power it did have lots of torque. Now even that is gone. The redline is at 6000 rpm but nothing is gained by pushing it above 4000. The engine always sounds like it's straining and wheezing, leading one driver to comment, "The engine performance makes the car feel like it weighs nine tons." It was no surprise in tallying the scores to discover the MGB mired in last in the engine category.

Even though it doesn't go and it doesn't handle, the MGB is still a reasonably comfortable car. The seats are comfy, the pedals are well positioned and the instruments are readable and the steering wheel and gear shift lever fall easily to hand.

But these virtues can hardly compensate for the antique qualities of the rest of the car. Perhaps British Leyland will have the last laugh. The MGB could be the first replicar that never went out of production!

MG Midget

THEN WE come to the Midget. Fun to drive? Maybe. But a modern car, hardly. The Midget scored as many worsts as the X1/9 did bests, 13, and in no category did it score higher than fifth. That about sums it up: a seriously outdated automobile in striking contrast to the throughly modern X1/9.

EASTWARD HO!

What it was really like out there

BY HENRY N. MANNEY III

CONSTANT READERS of our magazine (advt.) may brood from time to time on how it is all a piece of cake, getting to drive the latest goodies and getting your picture took pointing at a hubcap or even Sally's knees. Actually it is terribly difficult, partially because one has to think of something different to say about a string of similar automobiles and partially because actually setting impressions (if any) down on paper is like fighting pterodactyls with a flyswat. Consequently all hands enjoy a day or two away from the Office even when said days are interlarded with business as who knows? an idea might spring out.

So I was happy when Tony called and said wot abaht a trip to the desert. Probably because my sister took me to all the Baghdad oaters at the Bayne when I was little, the desert has always seemed a romantic place with a burnoose behind every bush and stripped tents full of swaying harem girls just over the next rise. These days the Arab camel troops are busy peddling oil and Aladdin's Roc has metamorphosed into the sheriff's helicopter but tant pis, the air is still clear and panoche mountains lie on the horizon. Anyway the desert trip wasn't a mission to rescue Maria Montez for the 85th time but to compare six sports cars in open country, not from one set of stoplights on to the next. Without overlapping too much on the article next door, it can now be divulged that the cars were MG Midget, MGB, Triumph Spitfire, Triumph TR7, Fiat X1/9 and Fiat 124, all open (or opening) except the TR7 which meant that Dottie had to pass among us with sunburn cream. The principals shall remain nameless so that you may have the pleasure of trying to guess who's who from the snaps. Likewise you may guess why (1) the choice of cars was England vs Italy (2) why it was more like a Vintage run with two exceptions, but I can only assume that these were the only ones that could be got ready in time . . . and at that it was a near thing. Never mind, the air is clear etc.

We duly started from the Engineering Ed's house in Dana Point, laden down to the Plimsolls with waffles and ice cream (for breakfast?) and topped up the tanks at the foot of Ortega Highway which goes over a pass to Lake Elsinore. Nice place to get worked in and scrub in the tires, get the elbow joints loosened up, the old racing reflexes going and so forth but unfortunately we stopped three times between Capistrano and the summit so oil temp needles never came off the peg. The drill was to change cars as often as feasible, in order for eight people to get driving impressions of six cars (the camera car didn't count) and at each stop to make a host of notes. In addition we had some sort of timed contest to take down tops, resulting in three blue fingernails (no prize for guessing which top), one blood blister and a splinter, while there was also a photographic orgy befitting the old *LIFE* magazine. After that we scuttled down the hill through Elsinore and changing seats again, went off to lunch near Temecula. In this period I had the Spitfire, which at the back felt fine but at the front was early Allard, the Midget which nobody else seemed to want, and the Fiat twincam, a lovely car but one requiring the Ultra Farina driving position.

After a sit-down lunch we beetled off

The Midget's handling is, well . . . interesting and entertaining. There's an excessive amount of rear-wheel steering so the car darts around a lot. The steering is very quick but a little numb and the driver is constantly making little corrections and overcorrections to keep the car headed approximately in the right direction. The short wheelbase and the British cart springs cause the car to hop about even over small road irregularities. Blame the bumpers for part of the Midget's idiosynchrasies; British Leyland chose the same "solution" for the Midget as it did for the MGB.

The brakes, however, were better than on the previous Midget we tested. They still fade excessively and the panic stopping distances are long but at least the control was very good.

This latest Midget uses the Spitfire engine, but to fit the narrow confines of the Midget's engine compartment the exhaust manifolding had to be made more restrictive so in Midget trim it produces two less horsepower. Like the Spitfire, the Midget was a 49-state car requiring premium fuel for its 9.0:1 compression ratio ohv engine. As might be expected from two cars differing in curb weight by only 50 lb the straightline performance is virtually identical up to around 70 mph when the Spitfire's better breathing and better aerodynamics afford it an advantage.

The Midget is aptly named for inside it's a hopelessly cramped car. This didn't seem to bother one driver who commented, "Once inside it isn't too bad. It gives me the impression of wearing the car, which lends a nice touch and feel to it all." Others were less kind, complaining about lack of seat travel, limited shoulder room and awkwardly positioned door handles. We all agreed on one point, however. Anyone who would buy a Midget when the X1/9 was available for only $1000 more would have to be a complete masochist.

Conclusion

IF THERE'S one thing this test proves conclusively it's that time and automotive engineering stand still for no car. A car that might have been perfectly adequate and more than acceptable 10 to 15 years ago when stacked up against its peers, can't hope to compete against a design that has benefited from the latest in automotive innovations and technology.

Thus we find the thoroughly up-to-date X1/9 and TR7 leading this comparison test. The Fiat 124 is one car that seems to disprove the above rule, but then the Italians always seem to design sports cars with more finesse than other manufacturers and there's no denying that with the passing of time it too is becoming dated. There's little hope for the three older British cars despite how much we enjoyed the Spitfire. The only thing left is for British Leyland to start all over with a clean sheet of paper.

---

in the general direction of Borrego Hot Springs along nice windey roads, doing our musical chairs routine at what seemed like 10-minute intervals. We didn't necessarily travel in a crocodile as at each stop there would be a certain amount of discussion about who had "had" what and those who got shed of discussion or note taking first shot off in a cloud of dust to wait at the next stop, duly marked out on the map by Old Wagonmaster Dinkel. The hot shots of course would go thundering past at a high rate of knots while those who enjoyed the scenery (translation: row it along with the gear lever) trundled along smiling broadly like those models in TV commercials. Someplace in here I was lucky enough to snaffle the Fiat X1/9, which I had avoided before as looking like a tichy little mock Can-Am, and straightaway found myself an instant Jackie Stewart. The wedge shape must give an effect like that of a wing on racing cars as the little devil was fairly glued to the road. It was a bit of a struggle to get into, nothing compared to the Midget, but the driving position was perfect for me and the power wasn't that bad. So this is a modern sports car. Land o' Goshen.

Toward the end of the day we found ourselves on the descent to the desert proper, with the Salton Sea glinting in the distance, and found a nice cowboy movie/Indian ambush valley to do sexy sunset photos in, all lined up among the puckerbushes. After that there was a Gadarene rush down the hill to Borrego with yr fthfl svnt in the MGB marginally leading the Engineering Ed in the Fiat 124. It was grand with the tops down, air temp just right and a nice wide road with no actual hairpins but plenty of tight sweepers. To be frank the MGB wasn't running really well, feeling as if the tappets were all too tight (a quart or two surplus in the sump didn't help), but downhill it was both comfortable and stable, sliding through corners majestically like a Lago Talbot and not caring much whether the brakes or throttle were off or on. Dinkel claims he cooked the front brakes on the Fiat which just proves, as I never heard of such a thing before, how much of a chicane I can really be.

We stopped the night at the Casa de Zorro, a posh motel out in plein boonie, and were assigned a couple of nice cottages with two bedrooms apiece. Fear and confusion arose immediately as one room in each had a double bed and (1) none of the men wanted to sleep together (2) we wouldn't have minded accommodating the girls but most of us were already married to somebody else. So that was straightened out by changing cottages, we all got gussied up and then went off to a hilarious dinner which was v good if a bit pricey. The nicest thing about the desert is that there are absolutely millions of stars out at night.

An unwilling crew turned out in the morning for brekkers (porridge in evidence) and after a certain amount of checking oil levels roared off up the road to Julian with camera car ahead, only to stop right away in the nice sun for more pix. I went to sleep. Then it was continuing on through the desert with occasional car changes, wondering what a horror this waterless waste must have been to the Butterfield Stage people. I had the Midget along some of the straight but bumpy roads along here and the rear-end hoist to meet Fed bumper requirements doesn't sit well with Midget rear suspension, which was marginal anyway; it was easy to imagine oneself as Nuvolari in the big Auto Union, sitting way up front with the back end steering him about. However the twisting pass up to Julian was more fun, flat out in 3rd about 4500, as the Sprite, sorry Midget, would get itself over on the bump stops around corners (one side quicker than the other as there was a list to port, at first attributed to the pancakes) and then you could get away with murder. Water temp stayed good which is an improvement from the bad old days.

After a stop in Julian, we trickled off toward Lake Henshaw as that road looked promising; while on the way we had the usual fun with locals in pickups and playing tag with the camera Rabbit for photos. The run down towards Pala Mission through the avocado groves was rather fun, especially since I had the TR7 for the first time. This has the engine similar to my Saab one and is quite smooth but seemed to be less revvy and down on power, probably because of its newness. For a conventionally suspended car it really keeps the road quite well, although I think that both Fiat and Alfa handle that better, but Triumphs have done a lot of homework. The 7 was awkward to get in and out of and I was never really comfortable, even though it was way ahead of the other three English models on modern feel.

Well, that was about it, objectively speaking, as traffic got worse as we approached the coast. As you can see it is all very taxing and fraught with responsibility, trying to hurry a car with built-in unkorrektheid, especially when occupying both lanes of a downhill road (in company with the camera car) and finding oneself face to face with a rusty Bel-Air. Seriously, though, little jaunts like this are the only way you can actually find out if the car can hack it, short of going to Europe of course (are you listening Tony?), and we hope that you will profit thereby.

Salon: THE LEGENDARY R1

Ken Miles' original MG Special runs, and wins, again

BY JAMES T. CROW

PHOTOS BY JOHN LAMM

For those of us who were around western sports car racing in the early Fifties, R1, Ken Miles' first MG Special, was of extraordinary significance. No other car of its time was so consistently outstanding, gave us as much pleasure or was so close to our hearts. That car now has been restored by its current owner, Nels Miller of Culver City, California, so you better brace yourself for some heavy-duty reminiscing.

First, the scene. It was early 1953 and at a sports car race meet there were usually two main events, one for modified cars under 1500-cc displacement, the other for modifieds over 1500. Earlier, the under-1500 race had been dominated by Roger Barlow's 1.3-liter Simca Specials. Then came the 1342-cc Oscas and the early Porsche Supers. These wiped up the Simcas and left only crumbs for everyone else. It was time for R1 to make its appearance.

These photos and the 2-page photograph were shot on the original start-finish straightaway of the Pebble Beach race course, with the polo fields as a background.

R1 made its competition debut at Pebble Beach that spring. The driver was Ken Miles, a blade-faced Englishman who was Service Manager for Gough Industries, the southern California MG distributor. In the under-1500 feature for the Pebble Beach Cup there were not only Porsches but three of the dreaded Oscas. Miles' new car had been impressive in practice but because this was its first race nobody seriously expected it to last the distance. It did, though, the victory was his and it was marvelous.

For the next year Ken and his dark green MG Special defeated everything and everybody in the under-1500 category. They won at Golden Gate, Phoenix, Madera, Chino, March Field, Palm Springs and Bakersfield. In addition to the under-1500 main events and Saturday preliminaries, R1 usually also ran in the over-1500 main as well, always beating a host of bigger cars and sometimes finishing as high as 3rd overall.

In its 10 main-event starts with Ken driving, R1 was never beaten. Eight times it won and twice it failed to finish, once at Moffatt Field with a broken clutch, then at the 1954 Pebble Beach with the flywheel bolts sheared. This Pebble Beach race, which it was leading at the time, was its last race with Miles at the wheel. Shortly afterward, with the construction of his new special underway, he sold it to another southern California driver, Cy Yedor.

Many specials seem to be 1-man cars; they work for the original owner-builder-driver but never do well for subsequent owners. Not so with R1. With it, Cy Yedor did well in southern California racing, generally being found in 2nd place behind Miles' new special, R2, the Flying Shingle. When the Shingle wasn't on hand, R1 was still capable of winning, as it did at Torrey Pines in late 1954 and at Hansen Dam in 1955.

Yedor campaigned R1 through 1954 and most of 1955, then sold it to Norris "Dusty" Miller, a southern California contractor. Dusty wasn't as skilled a driver as Miles or Yedor but he not only took lower-echelon trophies with fair regularity, he won an SCCA Regional race with it at Hourglass Field in 1956.

When Dusty moved on to a 1.5-liter Maserati in late 1956, R1 went unused until 1958 when it was entered in a hillclimb at Willow Springs. It was there that Dusty's son Nels raced it for the first time. Nels wasn't old enough to hold a competition license then, however, so the car went back into retirement. In 1961, when Nels became 21, Dusty gave it to him as a birthday present and it began a new competition career.

Now R1 was raced in Formula 2, which was for 1500-cc cars at that time. Running without fenders or headlights, painted yellow and with a fiberglass nose to replace the battered original, R1 was almost unrecognizable. It remained true to its pedigree, though, winning F2 with regularity and continuing its record of having won for each of its four owners during a career that spanned nine years.

When Nels moved on to other racing cars, R1 went back into

retirement, neglected but still basically intact. After its restoration its first public appearance was at the Monterey Historic Car Races at Laguna Seca in August 1976. It is wholly appropriate that it carried Nels to overall victory in the race for cars of its era.

Nuts & Bolts

OTHER ENTHUSIASTIC owner-drivers built MG Specials in those years and some of them knew what they were doing. None of those cars, however, had that combination of speed and reliability that set Miles' car apart. What made that car so different?

R1 is a legendary car and like most legends there is more than one version. The folktale that grew up around Miles and the car said R1 was put together of mostly stock parts by Miles and a few friends using nothing more sophisticated than basic hand tools and a welding torch. The other version, which has had less publicity, is that it was constructed as a Gough Industries project in order to sell MGs.

The car was designed by Ken and it was based on his experiences with a variety of specials back in England. John Beazley, Ken's boss at Gough, insisted that as many stock parts as possible be used to give the car identity as an MG. He also rode herd on the project to keep it within practical bounds since he felt Ken was inclined to go to greater lengths and more complicated solutions than were actually required within the rather simple concept. The fabrication was done by two Norwegian mechanics who worked for Gough, Lorenz Melvold and Arne Bjerkli, and the construction took place in the Gough service department on Barranca St in the Lincoln Heights section of Los Angeles.

The car had a simple but effective ladder-type frame, with main rails of 3.5-in. mild steel tubing. The part of the legend that says the car used many components straight from the parts bins is literally true. The torsion-bar front suspension and rack-and-pinion steering came from a Morris Minor, as did the radiator. The gearbox was stock MG TD, the rear axle TC. The front suspension arms used TD geometry but were fabricated of 1-in. steel tubing and the hubs were modified MG TC. The brakes were basically MG TC but the fronts were modified to two leading shoes and Alfin (aluminum-finned) drums were used all around. The rear suspension was by quarter elliptic springs below and radius arms above the solid rear axle and the shock absorbers, which were inclined forward at 45 degrees, were modified Monroe.

The engine was the only component not readily available to the everyday customer. The original plan had been to use a short-stroke Laystall crank, reduce the displacement to just under 1100 cc and then supercharge it, this being the formula for a supercharged car at that time if it were to run in the under-1500 class. Before the car was complete, however, the MG factory made available one of the experimental engines built for EX179, a streamliner record car driven by Goldie Gardner to many international records. This engine had one of a small number of special blocks in which the bore was 72.0 mm instead of the stock TD's 66.5, giving a displacement of 1466 cc instead of 1250. The horsepower figure given by Ken in a *Road & Track* story about the car in December 1953 was 81 at 6300 rpm.

There is a countering version of the legend that says this engine, while it may have started as little more than a meticulously balanced 1.5-liter version of the Mark II engine, was continuously modified until its compression was astronomical and its output well over 100 bhp.

Nels Miller, who has been through the engine more than once, confirms that virtually all the basic pieces, including crank and rods, are pure MG. Nels is also an experienced driver and he estimates its output at no more than 85 bhp, certainly not 100 or more. The cylinder head of the engine now in the car is stamped EX179/2, indicating that this is indeed the original engine. It is conceivable, however, that there was a second, more highly modified engine built up later and that it was transplanted to R2, the Shingle, while this original engine was put back into R1 before it was sold to Yedor.

The body of R1 was as simple as possible. Aluminum sheet was wrapped around the essentials and only the nose, where there were compound curves, had to be made by professional metal shapers. The body made little concession to aerodynamics but the frontal area was as small as possible, and it was probably a good solution to the problem considering the weight savings resulting from the simple bodywork. Ready to race but without fluids, Miles said it weighed 1225 lb. Nels Miller thinks that may be a bit heavy, estimating the weight at 1150 lb. Its light weight was obviously one of its great advantages, however, since the Porsche Super of the time weighed about 1675 and the Osca roughly 1650.

For the first race a section of wire mesh from a window in the Gough service department was used in front of the radiator but later a Morris Minor grille was cut in half and turned sideways in the opening. Engine-turned aluminum gave the instrument panel a nice touch and on the aluminum bulkhead behind the seats there was a small engraved plate with an MG emblem that read, "Special. R1. Designed by: Ken Miles. Built by: Lorenz Melvold."

The car's reliability was as astounding as its speed. Almost no modified car finished eight main events out of ten, especially ones capable of winning. R1 came to the track ready to race and it was seldom when anything more serious than a plug check was carried out in the pit or paddock. Folktales have it that this resulted from the simplicity, the use of proven components and the light weight.

Another explanation of its reliability gives credit to a stringent maintenance program, carried out primarily by mechanic Gordon Whitby. After each race there was a tear-down inspection of the drivetrain in which such items as clutch, axle shafts, wheel bearings and other frangible parts were routinely replaced. The

Below: Although campaigned without mercy by Dusty Miller, R1 continued to be one of the most reliable cars in its class. Bottom: R1 meets Osca. The more powerful Italian car dominated 1500-cc class until lighter, smaller, quicker R1 came along.

engine also came out of the car, was torn down, inspected and rebuilt before the next meet. The MG distributor had a large stake in the success of R1, this part of the legend says, so the greatest possible care was taken so the car wouldn't suffer any mechanical failure which would reflect adversely on the marque.

Maintenance like that may have been routine when Miles was racing the car and it also had good care while Yedor owned it since he had Bud Hand and Lorenz Melvold taking care of it. After Dusty Miller took it over, though, it ran until something broke, then was fixed and ran until it broke again. And it ran, mostly, usually failing to finish only when Dusty overrevved the engine and damaged the valve train. When Nels campaigned it he periodically checked the axle shafts, which did require periodic replacement, but he had neither the engine nor the gearbox apart during the year he raced it. And, as he points out, he was a wild boy-racer at that time so he wasn't at all gentle with the car.

Fit & Feel

NELS MILLER is taller than Ken Miles was and he doesn't find R1 especially comfortable to drive. His thighs bump the underside of the instrument panel, the steering wheel is too far away and there are various structural members that press uncomfortably against the flesh.

Driving it again after an interval of 15 years, though, he was reminded how much fun racing cars used to be. "It is a totally forgiving car," he says, "incredibly easy to drive."

The acceleration, compared to that of racing cars now, has to be ranked as modest and its top speed was probably never more than a little over 100. And the cornering power, which seemed so marvelous then, is not even on the same scale as that which has become commonplace with superwide modern racing tires.

Describing the car's handling, Nels says, "It's tame, compared to today's racing cars, but it's a marvelous car to drive. The steering is fingertip light all the way and the brakes are excellent. Even when I was racing it seriously, I never faded the brakes. It is a beautifully balanced car, absolutely controllable under all conditions, with final oversteer. You can drift it, slide it, do anything you want with it. You'd have to be incompetent to spin it out."

R1 always looked like a car that had absolutely impeccable handling. It's nice to have that confirmed.

Restoration

WHEN NELS removed the fenders and headlights back at the start of his racing career in 1961, he saved the brackets, realizing even then that he'd be restoring R1 someday.

In the years after that he became a serious racer, driving everything from BMW and Cougar Trans-Am cars to NASCAR Chevy stockers. He also worked in every phase of race car preparation and maintenance, becoming expert in everything from basic welding to engine building. As a result, when he began the restoration of R1 he brought to the job a proper appreciation of the car and do-it-yourself expertise of a high order.

The first step was to strip the car right down to the bare frame and this was done in January 1976. The rollbar that had been added later had to be removed, a few tubes replaced and some cracks welded but little else was required before the frame was sandblasted and repainted. He rebuilt the front suspension, duplicating each piece, using a fixture he'd made years before while building stock cars for Bill Stroppe. Every piece that came off the car was returned to as-new condition, or duplicated if it was too far gone to be restored. The nose was reconstructed by

Far left: R1 as it was raced as a 1500-cc Formula 2 car by Nels Miller.
At right: Ken Miles.
Bottom: Practice before Pebble Beach 1953, R1's first appearance on any track. Wire mesh in nose was later replaced by Morris Minor grille.

Dick Troutman, an old friend of Ken's, and the cycle fenders made by Red's Metal Shaping. Nel's lovely wife Jeannie fitted and stitched new leather around the original seat padding, Bud Hand furnished moral support as well as several bits and pieces, Brad Hand rebuilt the carburetors and a number of friends helped in various ways but it was Nels whose loving hands did the majority of the work.

By mid-July 1976, all the pieces were ready and the car was re-assembled with the Monterey Historic Car Races in late August as the target date. As it appeared at that event it was not completely finished. Nels had not yet found a Morris Minor grille of the correct vintage, the wood-rimmed steering wheel was slightly dished instead of flat and the numbers were stick-ons of a modern design rather than the somewhat crudely painted numbers it ran with when Miles raced it. Visually it also needed a British Racing Drivers Club decal on the cowl and of course it never ran with a license plate on the rear, the personalized "MG R1" plate being a nice but non-original touch added by Nels. He now has found a grille of the proper vintage, by the way, so R1 should satisfy even the most fastidious at its next appearance.

As for the felicity with which it was restored mechanically, Nels regards it as a restored *racing* car, meant to be raced, not just displayed. As a result he has felt free to use such things as modern high-pressure hoses for the external oil lines and replace the SU fuel pumps with something a bit more reliable. The tires also are not quite right. On the car now are 5.90-15 Dunlop Gold Cup tires when, originally, R1 used Dunlop Road Speeds, 5.00-15 front and 5.50-15 rear.

But it *is* R1, make no mistake about that. All the major components are exactly and precisely those that were in the car when Miles first raced it and for anyone who ever saw it race in those days, it is certain to bring a lump of nostalgia to the throat.

What it Meant

W HILE IT is true that R1 had an outstanding competition record and was never defeated in the 10 main event starts in which it was campaigned by the builder-driver, this alone cannot account for the car's impact or the reverence in which the car-driver combination was held. Ken Miles was a hero to us, a genuine hero driver of a stature no other driver quite enjoyed.

Being the builder-driver had something to do with it. Phil Hill was good, for instance, we knew he could win if he had a good enough car. But Phil didn't build his own car and then drive it to victory the way Ken did.

But there was more to it than that, too. Explaining a folk hero is complicated and that's what Ken was, a folk hero.

Most of us who were into sports cars at that time had MGs. We loved MGs. MGs had guts. They'd die trying. And unfortunately that's exactly what they'd do if they tried to compete with the German Porsche or the Italian Osca. World War II wasn't as long ago or so completely forgotten as it is now, and the attitudes that had been shaped in us from 1939 to 1945 hadn't disappeared completely on V-E Day. German industry lay in rubble and now

these ugly little jelly-mold Porsches with their guttural exhaust notes were arrogantly laying waste to the best our valiant MGs could do. And the Oscas? Even worse. We had little respect for Italy. To us the Italian armed forces had been the Polish joke of World War II.

Admittedly, I can hold a grudge longer than most people but I can't help but believe there was an element of this in the welcome that Miles and R1 received. Good old Ken, the intrepid Englishman, ex-sergeant of tanks, was out there fighting his own Battle of Britain against the Axis powers. And winning! Winning on skill and pluck and know-how. Trouncing the effete Italians, the heavy-handed, humorless Germans. (That the leading Porsche driver of the time had a Von in his name made our vicarious triumphs all the sweeter.) Take that, Red Baron!

The car also fulfilled part of the American dream that makes heroes of those who build a contraption in their backyard that is superior to that produced in the finest factories. Ken showed us that if you were good enough you could take even the humblest components and mold them into a world-beating race car.

It was such a marvelous car to watch, too. Miles had a sense of drama. Actually it was more than a sense of drama, it was a mean streak. He not only liked to win, he loved to humiliate his opponents. One of his tactics was to position R1 directly behind his fastest competitor and hang there, lap after lap, harassing and bedeviling him. If he could pressure the other driver into doing something stupid, and he often did, he'd pass him with a cheery wave and then come past the pits wearing that familiar twisted grin and we'd go wild with glee. Good old Ken!

On other occasions, if unable to make his opponent beat himself, Ken would finally pass wherever it pleased him and then pull away rapidly, often extending his lead at the rate of 3 or 4 sec a lap, making it obvious that his pursuers were no competition at all.

Or if the competition were really tough, as James Simpson's Osca was at the famous East vs West race at March Field, he had another tactic. As *Road & Track* described it, "Miles drove a strategically planned race, harrying the Osca through the corners and staying within shooting range on the straights. On the next to last lap, Miles took advantage of the Osca's failing brakes to pour on the coal and take the lead in the last lap and at the last possible moment . . ."

How good was Ken, really? My evaluation is that he was the best driver of his time in under-1500-cc cars. In that category, either then or later, I never saw him beaten except by a driver in better equipment.

In big-engine cars he was extremely good but he did not have that edge of superiority that had been his in the under-1500 class. He always had an excellent sense of pace and this made him outstanding with the Cobra team and later with the Fords that won the Manufacturer's World Championship. The year he was killed testing a Ford at Riverside, 1966, he had co-driven Fords to victory at the Daytona 24-hour and the Sebring 12-hour and would have won the Le Mans 24-hour race except for the grossest sort of grandstanding on the part of Ford's racing team management.

And R1? How good was R1? First, it was faster than Ken ever admitted. He said later it was never as fast as a 1342-cc Osca or a well tuned Porsche Super but it obviously was. It was, in fact, a very close match for the newer 1453-cc Osca and the early Porsche 550, both of which had 110 bhp but were considerably heavier than Miles' car. Like its driver, R1 was the best there was.

A further measure of its excellence came when it was passed on to other drivers. What other car can you think of that was still capable of winning its class nine years after it was built? And with the original engine?

It was probably only to be expected that in its race at the Monterey Historic Car Races it was R1 that won overall.

And based on that performance it's apparent that R1, now 24 years old, is ready to start another racing career, one that involves blowing the doors off its competition in vintage car racing.

May it run—and win—forever.

MGB

Hardly as new as tomorrow but it still has its points

IF THE MGB were the only sports car in the world—or the only one you'd ever driven—what a wonderful car you'd think it was. The responsiveness. The quick, accurate steering. The crisp action of the gearbox. The sharp, sure brakes. What fun it is to drive with the wind on your cheeks, the shift lever in your hand, the exhaust note in your ear. What a thrill to experience a feeling of one-ness with an automobile for the first time.

Somebody else said it like this: "To drive an MG is sheer pleasure. This is no car for the average Joe looking for transportation. Only those who know and appreciate the fun of driving a car which responds to skillful handling will ever get to like an MG. To drive an MG like an old maid is a sacrilege—a slow corner is taken in 2nd gear, a fast uphill bend in 3rd, high speed curves become a demonstration of accurate steering. Because an MG responds to skillful driving, its owner takes pride in improving his driving, yet because of its impeccable handling qualities,

the MG is far less prone to get into trouble, is in fact far safer than any other car of the family type."

Great stuff, eh?

Unfortunately, the MGB is not the only sports car in the world. And it's not the first one we've ever driven. Making it even sadder, the paragraph of praise quoted above was written by John Bond in 1954 on the occasion of the test of the then-new MG-TF. 1954!

But it isn't 1954 anymore and it isn't possible to ignore the intervening 22 years. We can't forget, for instance, that in the period of 1950 to 1955 MG introduced the TD, the TD Mark II, the TF, the TF 1500 and then capped off that half-decade with the MGA. Then came variations on the A, including a coupe, a Twin Cam and a 1600 before, in 1962, the MGB made its debut.

The B has been with us ever since. There was a body variation, the MGB-GT, a handsome variation on the sportswagon theme, but it isn't with us anymore since there was no economical way to make it pass the roof-strength requirements established by the feds.

In our first test of the MGB back in 1962, it had a 1798-cc inline 4-cyl engine developing 94 horsepower at 5500 rpm, went from zero to 60 mph in 12.5 seconds and scatted through the standing 1/4-mile in a highly respectable 18.5.

Our current test B also has the 1798-cc engine but it now has 63 bhp, takes 18.3 seconds to get to 60 and a leisurely 21.5 to cover the standing quarter.

By anybody's standards, this is modest performance. It is, in fact, almost exactly the same as the MG-TF tested in 1954—22 years ago. And that was the original 1250-cc TF, not the TF 1500.

The TF 1500 of 1954 would blow the 1976 MGB into the weeds!

Making it all even sadder, the B has retrogressed in other ways as well. To meet the U.S. bumper and headlight-height standards, the B has been cranked up on its suspension by what appears to be about three inches. This raises the body so it looks awkward and you can imagine what it does to the center of gravity. On the skid pad the current model lapped the 200-ft circle with a lateral acceleration of 0.698g while leaping, bounding and doing all sorts of ungraceful things. The MGB does not make a very distinguished demonstration of the way a sports car should handle on a constant-radius turn.

Can we think of anything good to say about the MGB? Of course. And it isn't all that difficult. The MGB has lots of virtues. It's great fun to drive, for instance, and in no way can be accused of being one of those dull, appliance-type automobiles. Helping make it fun to drive, the controls are where they ought to be and work the way they should. The shift lever, for instance, rises from the tunnel precisely where it should and snicks from gear to gear with a feeling that has always reminded us of the bolt action in a large-caliber rifle. Solid. Positive. Klatch.

The steering is also just right, having the correct balance between effort and results, making it second nature to position the vehicle exactly where you want it whether scudding along a winding road or easing into a parking place. The controls seem to be in tune not only with each other but also with the driver and after only a few miles you find yourself shifting to the correct gear without having to think about it and making just the right amount of steering input to track around a corner without winding and unwinding the wheel to correct and recorrect your original impulse.

There is a fabric top for the MGB but you should put this up only when rain or snow is actually falling. We don't mean that it's a bad top (admittedly it does restrict the headroom for the taller driver) but that the B is the kind of car you should drive with the top down. Figure on car coats, driving caps and string-back gloves as part of the investment in essential equipment. It will pay dividends in driving pleasure.

There's a saying among road testers about the good cars making you feel at home. In other words, one of the marks of a good car is how quickly you become used to driving it. The B ranks right up there with the best in this respect. On very short acquaintance you and the B begin to develop that feeling of oneness where the reasonably skilled driver feels confident in using it to its limits. Don't do anything stupid and it won't do anything to embarrass you. It's admittedly as old-fashioned as a quill pen but dammit, quill pens have their place in this world too.

MGB

PRICE
List price, any POE $4750
Price as tested $5220

GENERAL
Curb weight, lb 2320
Test weight 2650
Weight distribution (with driver)
   front/rear, % 51/49
Wheelbase, in 91.1
Track, front/rear 49.0/49.3
Length 158.3
Width 54.9
Height 50.9
Ground clearance 4.2
Overhang, front/rear 29.6/38.1
Usable trunk space, cu ft 2.9
Fuel capacity, U.S. gal 14.0

ENGINE
Type .... ohv inline 4
Bore x stroke, mm ...... 80.3 x 89.0
  Equivalent in.............3.16 x 3.50
Displacement, cc/cu in....1798/110
Compression ratio 8.0:1
Bhp @ rpm, net .... 62.5 @ 5000
  Equivalent mph.......................91
Torque @ rpm, lb-ft ... 72 @ 5000
Carburetion ...... 1 Zenith-Stromberg
Fuel requirement.... 91-oct unleaded

DRIVETRAIN
Transmission ....4-sp manual
Gear ratios: 4th (1.00) 3.91:1
    3rd (1.38) 5.40:1
    2nd (2.17) 8.48:1
    1st (3.44) 13.45:1
Final drive ratio 3.91:1

CHASSIS & BODY
Layout .... front engine,
  rear drive
Body/frame .... unit steel
Brake system: 10.8-in. discs front,
  10.0 x 1.7-in. drums rear
Brake swept area, sq in 310
Wheels .... 14 x 5½
Tires 165SR14
Steering type ....rack & pinion
Overall ratio 21.4:1
Turns, lock-to-lock 2.9
Turning circle, ft 32.0

Front suspension: unequal-length A-arms, coil springs, lever shocks, anti-roll bar
Rear suspension: live axle on leaf springs, lever shocks

INSTRUMENTATION
Instruments: 120-mph speedo, 7000-rpm tach, odo, coolant temp, oil pressure, fuel level

ROAD TEST RESULTS

ACCELERATION
Time to distance, sec:
  Standing ¼-mi, sec............. 21.5
  Speed at end, mph.................65
Time to speed, sec:
  0–30 mph 5.5
  0–50 mph 12.8
  0–60 mph 18.3
  0–70 mph 26.5
  0–80 mph 39.0

SPEED IN GEARS
4th gear (4950 rpm).................. 90
3rd (6000).............................79
2nd (6000) 51
1st (6000) 32

FUEL ECONOMY
Normal driving, mpg 22
Cruising range (1-gal. res)..285

HANDLING
Speed on 100-ft radius, mph .. 32.5
Lateral acceleration, g 0.698

BRAKES
Minimum stopping distances, ft:
  From 60 mph 176
  From 80 mph 320
Overall brake rating ....fair

SPEEDOMETER ERROR
30 mph indicated is actually 31
50 mph 53
60 mph 64
70 mph 74
80 mph 84

Salon J4 MG MIDGET

Retelling the tales and tasting the flavor of how it used to be

BY JAMES T. CROW

BUILT IN 1933, the J4 Midget came along when, as the purists say, MGs were still MGs. That was when the MG Car Company was still owned by the Morris Garages and before it became part of Morris Motors Ltd in 1935, which soon led to the decree that took the factory out of racing. But that's getting ahead of the story . . .

The name, MG, as everyone must know by now, was taken from the parent company, the Morris Garages of Oxford, owned by Sir William Morris, later Lord Nuffield. The Morris Garages came along early in the creation of Morris' empire and, in fact, were established to sell and service cars and motorcycles several years before he began to manufacture the cars that bore his name.

In 1922, Morris hired Cecil Kimber as general manager of the Morris Garages. Kimber was an enthusiast, which Sir William was not, and one of his innovations was to have special bodies built onto Morris chassis. The first of these was the "Chummy," which was grafted onto the workaday Morris Cowley and the result was a somewhat sporty roadster with a mother-in-law seat in the rear.

The first true MG, later famous as "Old Number One," was neither the first car produced by the company nor the first to bear the MG name, since this was already being used on the rebodied Morrises by this time. Old Number One does qualify as the "first true" MG, however, since it was the first MG specifically designed for a sporting event, the Lands End Trials of 1925. Kimber himself drove it there, completing the entire course without penalty points and thereby qualifying for a gold medal.

Although Kimber's special was an appealing machine, the company did not immediately turn to the production of a competition car, instead continuing with the construction of sedans and coupes on various Morris chassis. In the spring of 1928, the MG Car Company was officially created as an entity

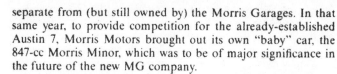

separate from (but still owned by) the Morris Garages. In that same year, to provide competition for the already-established Austin 7, Morris Motors brought out its own "baby" car, the 847-cc Morris Minor, which was to be of major significance in the future of the new MG company.

By altering the frame and suspension of the Minor and adding a sporty 2-seater body, Kimber created the first of the MG Midgets, the Type M. Although fraught with limitations attendant upon its undistinguished heritage, the M-type was popular with club racers of the day and competed with honor in speed trials, rallies and even endurance events such as the Le Mans 24-hour race and the Double 12-hour at the Brooklands circuit near London. By the time it went out of production in mid-1932, more than 3000 M-types had been built and the Midget concept was firmly established as perhaps the single most important element in the company's line.

EX-120 was the next high-performance MG of note. This was a special built for Capt George Eyston (pronounced Easton) which was based on the M-type and used an engine modified to conform to the international 750-cc displacement class. Using a Powerplus supercharger, Eyston set a host of speed records at the Montlhéry circuit near Paris and went more than 100 mph, the first 750-cc car to attain such a velocity.

This was followed by the Type C, a competition model known as the Montlhéry Midget and powered by the Morris Minor engine, destroked to 746 cc and available with or without supercharger. A less-expensive companion model using the 847-cc engine from the M-type was designated the D. The C- and D-types were built from mid-1931 until mid-1932 and were followed by the first of the J-type Midgets and the K-type Magnettes, perhaps the most famous and most successful of all the racing MGs.

The J-types

FIRST SHOWN in August 1932, the J-type set the style for all Midgets right up to the last of the line, the TF, in 1955. And a proper sports car it looked, too. It had the deeply cut doors formerly seen only on racing models, forward-folding windshield, double-hump cowling and the blunt tail with the upright fuel tank and rear-mounted spare tire. It was lean and lithe, that J-type, sitting on tall, narrow 4.50 x 19 tires on an 86-in. wheelbase and 42-in. track.

PHOTOS BY WM. A. MOTTA

Altogether there were four J-types. The J1 and J2, 4-seater and 2-seater, respectively, had the 4-cylinder, 847-cc engine derived from the Minor but significantly improved through a series of modifications that included a hotter camshaft and a cross-flow head with a pair of semi-downdraft SU carburetors. The J3 differed from the J2 in that it was equipped with the supercharged 746-cc engine and the J4 was the all-out racing version with lightweight body panels, no doors, a further-modified engine and brakes of 12-in. diameter where the other Js made do with 8s. The engine of the J4 was modified in several details, one of the most important being the use of a fully counterweighted billet crankshaft. This was significant because this intrinsically humble engine had but two main bearings, the rearmost of which tended to go away under hard use. With the counterweighted crankshaft, the J4's engine was rated at 71 bhp at 6000 rpm where the 847-cc version's maximum output of 36 bhp came at 5500 rpm.

The transmission was a 4-speed manual of the crash variety, meaning there were no synchronizers between gears, and "crash" was the noise it made when shifted without regard for matching engine speed to transmission speed.

There were solid axles at both ends and these were suspended on semi-elliptical springs that used sliding trunnions instead of conventional shackles to accommodate the change of length occasioned by the up-and-down movement of the axles. Not that there was a lot of up-and-down action, you understand, since generous suspension travel was not the style with sports cars of that time and any tendencies in that direction were further discouraged by friction-type shock dampers. The frame was of ladder-type construction with the side rails running underneath the axle at the rear but conventionally upswept at the front.

The J4 was a hairy beast to drive, the consensus being that it was too powerful for its chassis and there is good evidence that this opinion was justified. Only one driver, Hugh Hamilton, was consistently able to extract its best performance and it is "Hammy," as he was known, whose name is most often associated with the car's success. He got one of the first J4s to be produced, entered it in the Eifelrennen at the then-new Nürburgring circuit in Germany and not only broke the 750-cc lap

rewriting that particular page of motor racing history.

The Tourist Trophy was a handicap race in those days, each class being assigned a target speed based on displacement. The slowest classes started first and if the handicapping were perfect all the cars in the race would reach the finish line at precisely the same moment. Lacking such perfection, the winner was the car that beat its bogey time by the greatest amount.

A total of 25 cars ranging from eight 750-cc Midgets to a pair of 4.5-liter Invictas lined up for the start and were flagged off, class by class, rushing over the narrow country roads that made up the 13.7-mile circuit. The race would last almost six hours and

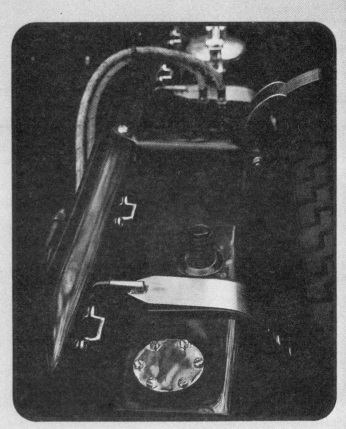

MG J4 MIDGET

Price, new	£495
Engine	sohc inline 4
Bore x stroke, mm	57.0 x 73.0
Displacement, cc	746
Bhp @ rpm	72 @ 6000
Carburetion: SU carburetor with Powerplus supercharger	
Clutch, type	dry
Transmission: non-synchronized 4-sp manual	
Gear ratios: 4th (1:1)	5.38:1
3rd (1.36)	7.31:1
2nd (1.86)	10.00:1
1st (2.69)	14.46:1
Final drive ratio, std	5.38:1
Wheelbase, in.	84.0
Track, front/rear	42.0/42.0
Overall length	124.0
Width	51.5
Height, top of cowl	42.3
Fuel capacity, U.S. gal.	12.0
Layout	front engine, rear drive
Frame/body: ladder-type frame underslung at rear, separate light alloy body with wooden substructure	
Brakes: 12.0-in. drums front and rear, mechanically operated	
Wheels	19 x 3
Tires	4.5 x 19
Front suspension: beam axle, semi-elliptics, Hartford shocks	
Rear suspension: live axle, semi-elliptics, Hartford shocks	
Instrumentation: tachometer, supercharger press., oil press., oil temp, water temp, fuel level, oil tank level, oil sump level, ammeter, clock	

record but also won his class by almost half an hour. He scored other triumphs in the J4 but his greatest day was to come in the 1933 RAC Tourist Trophy race at the Ards circuit near Belfast in Northern Ireland when he and the J4 challenged the immortal Tazio Nuvolari in the K3 Magnette.

The 6-cylinder K3 Magnette had won the 1100-cc class and the team prize at the classic Mille Miglia in Italy shortly before this and thus convinced the great Italian driver it was a machine worthy of his talents. Nuvolari's performance in the TT is one of the legends of motor sport but the story that is less often told is that Hammy and the J4 came within a clumsy pit stop of after the first of these Hamilton was leading by 53 seconds. There was considerable sorting out to be accomplished in the early going because of the differences in potential speeds of the cars, but after Brian Lewis' blown 2.3-liter Alfa Romeo retired with a broken transmission the race settled down to a duel between Nuvolari and Hamilton in the two MGs. Both cars made a mid-race pit stop and it was here that Hamilton lost the race. Nuvolari and his mechanic pulled off a near-perfect stop of just over 3 minutes and part of the legend tells how the cool Italian driver made one final circuit around the car, tapping each knock-off spinner with a hammer, making certain it was properly done up.

Hamilton and his mechanic, on the other hand, had everything go wrong. The starter jammed, the mechanic set himself on fire, they leaped aboard only to discover nobody had closed the hood, there were shouts, screams and curses, and before they were finally underway again, gravel spewing from the tires, they'd spent more than *7 minutes* in the pits. At the end of three hours they'd had a seemingly comfortable lead of almost 6½ minutes. Now they were behind by 47 sec.

The stage was now truly set for one of the great duels in racing

Incredibly, Hamilton squeezed even greater speed from the J4, stealing back a second here, two seconds there until, miraculously, he was in the lead again. With but two laps to go the J4's lead had been extended to 54 sec and the race was all but won. But then, on the penultimate lap, Hamilton had to bring the J4 in for a splash of additional fuel. This time the refueling went smoothly but Nuvolari, whose K3 was down to the final drops in the reserve tank, continued around the track to win the race by 40 sec.

That September day of 1933 was to be the J4's greatest race. Yes, it had its fair share of success but racing cars represent the highest state of the automotive art and consequently tend to go out of date quickly. Less than a year later the Q-type Midget was introduced and it, also 750 cc but equipped with a 3-bearing crankshaft and Zoller supercharger, had more than 110 bhp, rendering the J4 instantly obsolescent if not obsolete.

So that was the J4. A brilliant car in its class, the fastest 750-cc car of its day, a car with a legend that will be told as long as enthusiasts gather to retell the tales and taste the flavor of how it used to be.

history as the two drivers broke and re-broke their lap records, and it is easy to visualize the Flying Mantuan pounding on the side of his K3, urging it to go faster, while Hammy, hunched over the wheel, his face in a deep scowl, inexorably clawed away at the bigger car's lead. With two hours to go, Hamilton had closed the gap to 9 sec. In another half hour it was only 3. Then Hammy came into the Dundonald hairpin too fast, skated into the escape road with all four wheels locked up, turning the air blue with his curses as he got turned around and underway again. With but one hour left in the race, Nuvolari led by a seemingly insurmountable 39 sec.

J4005

THE J4 Midget shown in the accompanying photos belongs to Gerard J. Goguen of Needham, Massachusetts. A hardcore MG buff, his interest in the marque has resulted in a collection that includes 37 different MG models ranging from a very early 14/40 4-place open tourer to a recent MGB, perhaps the most nearly complete MG collection in this country. He is also the proprietor of Abingdon Spares Ltd ("the world's largest MG-TC, TD, TF parts specialists," according to the classified ad regularly found in R&T's "Market Place") and, for contrast, a professional musician now in his 26th season playing trumpet with the Boston Symphony and Boston Pops. He acquired his first MG in 1953, was an MG dealer for a short time and then, in the late Fifties, began buying and restoring older models. "People heard I had extra parts and that's how I started my business, Abingdon Spares. Pretty soon I was having new parts made and everything went on from there." His first complete restoration was in 1960 when he rebuilt a TD from the bare frame up.

His J4 is serial number J4005, the fifth car in the line that began with J4001 and ended with J4009. He got it from a friend, Gary Schonwald, who saw it advertised in the December 13, 1970, issue of the *New York Times* ("An extremely rare race car, J4, disassembled but restorable"), bought it and then resold it to Jerry a short time later.

Originally dark blue with black interior, it was built in 1933 and the list price was £495, or about $2500 prewar. The first owner was a wealthy young man named Luis Fontes and it was registered in Nottingham as TV8371. Fontes is known to have raced it at Brooklands, possibly at Donington Park as well, and he was also a participant in the famous Tourist Trophy race of 1933, where records show him to have retired during the first hour with engine trouble. No record of Fontes ever having finished a race with the car has been found and Jerry would welcome information from anyone who is able to help him fill in details of its early history.

By 1936, J4005 was owned by one J. Hewson who replaced the original Powerplus supercharger with a 10-psi Marshall blower and there is then a hiatus in its history until 1944 when an American, Gene Aucott of Philadelphia, came across it under a

Hugh Hamilton's J4 Midget roars through Comber in 1933 Tourist Trophy. Story of his race against Nuvolari's K3 Magnette is a classic.

pile of tires at the Valentis Motor Car Co of Glasgow. Aucott bought the car, drove it for a few days in Scotland, then shipped it to the U.S. in the *S/S William Tilghman*. It was listed on the ship's manifest as "ballast" and arrived in New York in February 1945.

Later that same year it was sold to Francis Grant of Chestnut Hill, Pennsylvania, and then, in November 1945, to Otto Linton of Devon, Pennsylvania. By this time it had been painted a dark red, there was a connecting rod missing from the engine and the supercharger was gone. Linton fitted it with P-type rods, installed a Marshall supercharger and in 1948 entered it in the first

Watkins Glen Grand Prix where it failed to finish. Linton also raced it in several very early events at Bridgehampton, New York, Thompson, Connecticut and Langhorn, Pennsylvania.

It went through a series of owners after Linton, becoming less and less original. By the time it came to Jerry it had an engine and radiator from a P-type Midget, aluminum cycle fenders and hood and a Roots-type supercharger in place of the Marshall installed by Linton.

The car waited in Jerry's shop for a period of time while parts were sought that would assure the most authentic restoration possible. Through his connections Jerry had amassed a considerable quantity of possibly useful spares, including two J2s, which were far less rare because more than 2000 of that model were built. The single most important acquisition was two boxes of spare parts that had been shipped in 1934 from the MG factory to driver Jacques Menier in Paris. Menier had entered a K3 and a J4 at Le Mans that year and those spares, probably intended for use in preparing the cars for the 24-hour race, possibly arrived too late to be used since the crates remained unopened for 41 years. Among the J4 spares were many little nuts and bolts and special washers, connecting rods, the pipe that goes from the supercharger to the engine and, most valuable of all, a brand new Powerplus 8 supercharger which gives J4005 the distinction of being the only J4 still in existence with the correct blower.

As no J4 engine was available, the engine for J4005 was made in the same way as the original, by modifying the 847-cc engine used in the J2. Jerry shipped a J2 engine to England for that purpose and the work was done under the supervision of Eric Tieche of A.E. Tieche and Co, Wembley, Middlesex.

Eric Tieche's son Colin, also a J4 owner, spent six weeks at Jerry's where he overhauled the transmission, assembled a rear axle out of the parts of several which had been collected and performed many details on the chassis. The body's wooden substructure was found to be in surprisingly good condition and these parts served as patterns for the new parts to be made. The oil tank and exhaust system, including the fishtail, were made by Colvin Gunn in England, using photographs that enabled him to duplicate everything exactly, right down to the size of the tubing and the spacing of the rivets.

The restoration was completed in June 1976 and the engine was first fired up at an MG club meeting for the aural delight of those assembled. The next day he had as passenger the club's special guest, Capt George Eyston, the former Land Speed Record holder whose name was long associated with MG record cars. In J4005's first formal appearance it was awarded first place in the vintage MG category at the Gathering of the Faithful Mark XXII. The car is completely operable and Jerry confesses to having driven it a total of perhaps 20 miles. ("That fishtail could break your eardrum.")

Of the nine J4s built, five are now known to be in existence. J4002, which was the Hugh Hamilton car, is owned by Colin Tieche, who assisted Jerry in the restoration of his car. J4004 is owned by Molly Coles, widow of the vintage racing car enthusiast Geoff Coles, and J4006, also owned by Coles at the time of his death, has been sold to a collector in Germany. J4007 belongs to Syd Beers, the British enthusiast who is the custodian of the world's largest collection of vintage MGs. The whereabouts of J4001, J4003, J4008 and J4009 are not known although it is believed that either J4003 or J4009 was destroyed in a bombing raid on Bristol during World War II.

J4005 is the only J4 in the U.S. and it is a source of pride with Jerry that it is also the most nearly original of the few that remain. His plans for the car include racing it at the vintage Sports Car Club's Lowenbrau Race at the Watkins Glen 30th anniversary meet next fall and later it will be prominently displayed in the museum of MGs Jerry plans to open. There are other MGs Jerry looks forward to adding to his collection ("I'd like an L-type, and a W- and a C-type") but it's also obvious from the superb restoration of the J4 and from the special glow that emanates from Jerry as he talks about it, that J4005 occupies a very special place in his heart. As well it should. As well it should.

MGB

We complain. British Leyland makes money. Time stands still.

AGING IS AN important (and inevitable) process in the scheme of things, especially with regard to fine cheeses, wines and people. In the arena of automotive design, however, aging can be a negative concept. Some car designs do age quite well and become classics, but others don't—the MGB seems to fall somewhere in between. This most modern (!) MG model is now nearly 17 years old and no successor is looming on the horizon. In fact, as British Leyland is pleased to note, MGB sales continue to run at healthy levels each year, so as long as the demand is constant, why spend the money to redesign or create a new model?

Well, that makes good business sense, but it's less than satisfactory for the enthusiast who would like to see an up-to-date sports car once again sporting the MG octagon, a marque badge that dates back to 1923, when it stood for Morris Garages.

The MGB was a nearly completely new car design when it went into production in June 1962, taking over from the MGA. The traditional MG ladder-type frame gave way to monocoque construction and while the inline, overhead-valve, 4-cylinder engine was basically a carryover from previous models, it was given an increased cylinder bore to bump displacement up from

1622 cc to 1798, where it remains to this day. While its driveability is reasonably good, aside from a reluctance to start when cold in our mild southern California climate, emission controls and a change in 1975 from two carburetors to one have reduced the engine's output from 94 SAE gross bhp (approximately 83 SAE net bhp) at 5500 rpm in 1962 to a current figure of 62.5 SAE net bhp at 4600 (the compression ratio is now 8.0:1 versus the original 8.8:1). Torque is currently 88 lb-ft at 2500 rpm compared to 107 at 3500 at the time of the B's introduction. Straight-line acceleration is not what it used to be, of course, but British Leyland recently made a helpful revision to the cooling system, replacing the engine-driven fan with a pair of electrically controlled ones, that has restored a small measure of the car's earlier performance. Our latest test revealed a 0–60 mph time of 13.9 seconds and the run through the standing-start quarter mile was accomplished in 19.8 sec at 69.0 mph. These figures are considerably better than those recorded in our last test of the MGB (June 1976); but a portion of the difference results from use of our new computer test equipment, which is operated by the driver alone and thus dispenses with the weight of an observer.

The engine is coupled to a 4-speed, fully synchronized manual

PHOTOS BY JOHN LAMM

This instrument panel is the third design used in the MGB since its introduction in 1962. The controls have been modernized along the way too.

gearbox that is crisp and precise in the best MG tradition. Our test model had the optional ($250) hydraulically actuated (via an electric switch on the top of the gearshift knob) overdrive, which has a ratio of 0.82:1. At or well above the present maximum speed limit in the U.S., cruising in 4th gear is pleasant and easy; the overdrive makes it just that much more so. The overdrive can be used with 3rd gear too, giving the driver increased flexibility in selecting ratios for various driving conditions, as when driving in the 25/45-mph range on city thoroughfares.

On the road, the MGB's handling characteristics are not those for which the car was once justifiably admired. The suspension design has remained fundamentally unchanged, comprised of unequal-length A-arms and coil springs at the front and a live axle on leaf springs at the rear. Admittedly not modern, this is not a hopeless layout; but MG chose to meet U.S. bumper-height requirements (new in 1974) by jacking the car up some 1.5 in. on the suspension! The result is a car that looks odd despite the handsome treatment of the bumpers themselves, defying the traditional notion that sports cars are low-slung, and one which does not progress through corners the way it should. On constant-radius turns, such as our skidpad, the MGB tends to leap rather than drive around the circle, and on the road there is more body roll than most drivers find comfortable combined with a feeling of awkwardness through medium to slow bends. Those who remember driving older MGBs can only lament what has become of a formerly good-handling sports roadster.

In matters of ride and noise, the MGB never was a comfortable or quiet car; sports cars were not expected to be in the days when it was conceived. But for those unfamiliar with the tradition it may need repeating here: Rattles, squeaks and gobs of wind noise are part and parcel of its character and must be accepted as such. Roadsters like this were always at their best with the top down; in this form all such unpleasantries are minimized.

The interior appointments have been downgraded and upgraded over the B's lifespan; today the car is livable for most drivers and passengers, though devoid of the leather seat facings it once had. We had some disagreement among our staff concerning the seats, with one group describing them as comfortable and pleasant while others felt there was considerable room for improvement, including one person who said he felt as though his lower back had been abandoned entirely.

Other criticisms centered around the seatbelts, whose guides were not firmly attached, allowing the belts to become easily tangled; and the handbrake lever, which is located between the

Seatbelt reels and their covers proved problematical on the test car.

PRICE
List price, all POE $5995
Price as tested $6590

GENERAL
Curb weight, lb 2335
Weight distribution (with
  driver), front/rear, % 52/48
Wheelbase, in. 91.1
Track, front/rear 49.0/49.2
Length 158.2
Width 59.9
Height 51.0
Fuel capacity, U.S. gal. 13.0

CHASSIS & BODY
Body/frame .... unit steel
Brake system .... 10.8-in. discs
  front, 10.0 x 1.7-in. drums rear;
  vacuum assisted
Wheels .... styled steel, 13 x 4½J
Tires .... Dunlop SP68, 165SR-14
Steering type .... rack & pinion
  Turns, lock-to-lock 2.9
Suspension, front/rear: unequal-
  length A-arms, coil springs, lever
  shocks, anti-roll bar/live axle on
  leaf springs, lever shocks

ENGINE & DRIVETRAIN
Type .... ohv inline 4
Bore x stroke, mm .... 80.3 x 89.0
Displacement, cc/cu in. .. 1798/110
Compression ratio 8.0:1
Bhp @ rpm, net ...... 62.5 @ 4600
Torque @ rpm, lb-ft ... 88 @ 2500
Fuel requirement .. unleaded, 91-oct
Transmission .... 4-sp manual
Gear ratios: 4th OD (0.82) .. 3.21:1
  4th (1.00) 3.91:1
  3rd OD (1.13) 4.42:1
  3rd (1.38) 5.40:1
  2nd (2.17) 8.48:1
  1st (3.44) 13.45:1
Final drive ratio 3.91:1

CALCULATED DATA
Lb/bhp (test weight) 40.1
Mph/1000 rpm (4th gear) 18.4
Engine revs/mi (60 mph) 3260
R&T steering index 0.93
Brake swept area, sq in./ton .. 248

ROAD TEST RESULTS

ACCELERATION
Time to distance, sec:
  0–100 ft 3.8
  0–500 ft 10.6
  0–1320 ft (¼ mi) 19.8
Speed at end of ¼ mi, mph 69.0
Time to speed, sec:
  0–30 mph 4.0
  0–50 mph 9.4
  0–60 mph 13.9
  0–70 mph 20.5
  0–80 mph 31.7

SPEEDS IN GEARS
4th gear (5000 rpm) 93
3rd (5500) 74
2nd (5500) 49
1st (5500) 32

FUEL ECONOMY
Normal driving, mpg 18.5

BRAKES
Minimum stopping distances, ft:
  From 60 mph 177
  From 80 mph 320
Control in panic stop .... fair
Pedal effort for 0.5g stop, lb 25
Fade: percent increase in pedal ef-
  fort to maintain 0.5g deceleration
  in 6 stops from 60 mph 60
Overall brake rating.................. fair

HANDLING
Speed on 100-ft radius, mph .. 32.5
Lateral acceleration, g........... 0.698
Speed thru 700-ft slalom, mph....53.0

INTERIOR NOISE
All noise readings in dBA:
Constant 30 mph...................... 70
  50 mph 75
  70 mph 85

SPEEDOMETER ERROR
30 mph indicated is actually .. 32.5
60 mph 61.0
70 mph 70.5

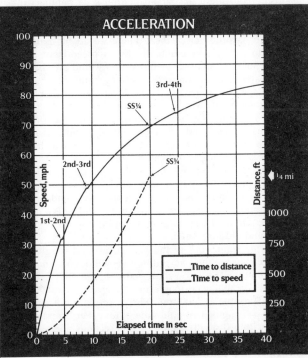

ACCELERATION

seats and frequently smacks occupants on the elbow as they enter or exit the car. The ventilation system is traditional British, i.e., lots of heat from the heater but precious little fresh air for cooling.

Well, that's why the top is convertible, right? Which leads us to the hassle involved in putting up or lowering the top. To a man and woman, every staffer who accomplished the feat came back with mixed feelings: pride at having done it, wonder as to why it has to be so complex. It requires an untoward amount of fiddling and pulling and even a little cursing to achieve the desired result.

Despite our criticisms of the MGB, it can be an entertaining sports car to drive if one keeps in mind its limitations. The rack-and-pinion steering is just right, for instance, with the proper balance of feedback and effort. Unfortunately, the brakes are only fair; stopping distances are rather lengthy and the brakes have a tendency to lock up more than we like. But a sunny day, an amiable companion and a winding country lane stretching before you can restore a lot of your enthusiasm for sports-car driving. The MGB still fits this scenario quite well. Despite its faults and limitations, and because of its virtues, many thousands of people keep driving and enjoying motoring as it used to be.

MG MIDGET

Why would anyone want one?

PHOTOS BY RON WAKEFIELD & JOE RUSZ

BACK TO BASICS is the phrase that most readily comes to mind in describing the MG Midget: It retains its unenviable position as the most basic sports car offered in the U.S. market. MG broke the barrier some 30 years ago and earned the affection of American sports-car enthusiasts, popularizing an entirely new genre of driving pleasure—small, lightweight, open 2-seater automobiles with an allure virtually unknown on this side of the Atlantic before then. There would be many other marques to follow and eventually MG's preeminent position would decline, but no other manufacturer would ever be able to take away the British carmaker's firstborn status.

The Midget was introduced in 1961 as an offshoot of the Austin-Healey Sprite Mk II, which had made its debut in Mk I ("Bugeye") form in 1958; thus the diminutive MG has one of the longest continuous production runs in the sports-car world. There have been many changes in the Midget since its birth, but it's still the same type of car it's always been: basic roadster. True, now there are windup windows, plush carpeting and other amenities scarcely dreamed of in the days of the original Midget, and U.S. safety and emission regulations have mandated many revisions and alterations, but a drive in the Midget is a nostalgic experience. Or, as the less charitable put it, the car is seriously outdated and old-fashioned.

Within the framework of the British Leyland conglomerate, the Midget shares the same inline 4-cylinder engine as the Triumph Spitfire. It's an overhead-valve design with a displacement of 1493 cc that in current form develops 50 bhp at 5000 rpm and 67 lb-ft torque at 2500 rpm. The powerplant is a workhorse "tractor-type" engine in a mild state of tune. Performance is not startling, with a 0–60 mph time of 14.3 seconds and a quarter-mile clocking of 20.3 sec at 69.5 mph (measured with our new lightweight test equipment). That 0–60 mph time is 1.2 sec

quicker than the Midget tested in June 1976, but the 1978 model is actually 0.2 sec slower in the quarter mile. Chalk that up to a reduction in compression ratio (9.0:1 down to 7.5:1) and horsepower (55.5 down to 50.0) since our 1976 test.

The Midget's 4-speed fully synchronized gearbox has a notchy feel and a high shift effort when new, but as it breaks in it improves greatly and is crisp and positive wihout being too stiff. The transmission is well matched to the engine's characteristics and makes good use of the low- and mid-range torque, enabling the Midget driver to keep pace easily with traffic around town.

A perusal of the Midget's dimensions reveals that it's aptly named, sitting on an 80.0-in. wheelbase and measuring just 143.0 in. overall. The short wheelbase, coupled with the essentially simple suspension design (lower A-arms, lever shocks as upper lateral arms, coil springs and anti-roll bar front; live axle on leaf springs with lever shocks rear), make for handling characteristics that are entertaining to some and nerve-jangling to others. One staff member described the Midget this way: "It can really be thrown about; the small size and quick reactions grow on you. The Midget's potential may be fairly low, but it's very easy to use all of that potential." Others were markedly less enthusiastic, finding the Midget something of a handful, including one person who described it in nautical terms as "very much like tacking my way up the boulevard," constantly nudging the steering back and forth to maintain a somewhat straight line. The driver who masters the Midget's handling idiosyncracies will be rewarded with some entertainment on a winding road with a smooth surface, but pushing to the limit brings about the onset of final oversteer, requiring quick driver reactions and a deft touch. The fast (2.3 turns lock-to-lock) rack-and-pinion steering also necessitates a light hand on the wheel and a bit of practice to keep from overdoing it when maneuvering.

The ride manners of the Midget are rather unpleasant. As a runabout, it's quite acceptable and the firm and jouncy nature of the car seems appropriate. For anything more than a short drive, however, the stiff-legged feel becomes tiresome.

Inside, the Midget can be described at best as cozy, at worst as claustrophobic. Perhaps it's a natural law of selection that this car is most suited to young and lithe people who don't mind contorting themselves a bit to get behind the wheel. Once there, though, they will find that leg, hip and shoulder room are restricted for anyone taller or more portly than average. The controls and instruments are well laid out and most everything (including the opposite door handle and window crank) is within easy reach—many of our staff members expressed the feeling that

it's more like wearing a car than driving one. There is insufficient space to move the seat far enough rearward for most drivers more than about 5 ft 8 in. tall; one short driver found that when the seat was close enough to depress the clutch pedal fully, she had great difficulty reaching back to use the door handle to get out! One quite serious problem concerns the safety belts—the inertia reels stick badly as you try to put on the belts, and when the convertible top is down it interferes with them so much that it becomes a major task to adjust and fit them. The reel covers won't stay on the reels, further interfering with one's good intentions to belt up. Also, the receptacles for the belt latches on the inboard side of the seats present a dangerous and hard surface to many people's hip bones. The convertible top provides

Current Midgets share their engine with the Triumph Spitfire.

MG Midget's cockpit is one of the tightest known to man.

a reasonably snug, closed environment but there is considerable wind leakage around the edges and the Midget is a car best reserved for a sunny day when the top can be left down and the driver is free to enjoy open-air motoring in the old style.

Putting the top up and down is easier than it used to be, but there are still far too many steps required, snaps to be fastened, etc. Other carmakers have found much simpler solutions to the convertible top (the Fiat 124 comes to mind) and there's no reason why British Leyland couldn't follow suit.

All in all, we find far more minuses than pluses with the MG Midget. Small, lightweight sports cars can and should be great fun to drive. Perhaps many people can and will experience that sort of fun with the Midget. But there is ever so much room for improvement, badly needed to return to MG the rightful crown as builder of exciting sports cars. As it stands, the marque has fallen into a state of disrepair.

PRICE

List price, all POE	$4850
Price as tested	$5260

GENERAL

Curb weight, lb	1835
Weight distribution (with driver), front/rear, %	52/48
Wheelbase, in.	80.0
Track, front/rear	46.3/44.8
Length	143.0
Width	54.0
Height	48.3
Fuel capacity, U.S. gal.	7.5

CHASSIS & BODY

Body/frame	unit steel
Brake system	8.3-in. discs front, 7.0 x 1.1-in. drums rear
Wheels	styled steel, 13 x 4½
Tires	Pirelli Cinturato CF67, 145SR-13
Steering type	rack & pinion
Turns, lock-to-lock	2.3
Suspension, front/rear: lower A-arms, lever shocks as upper lateral arms, coil springs, anti-roll bar/live axle on quarter-elliptic leaf springs, lever shocks	

ENGINE & DRIVETRAIN

Type	ohv inline 4
Bore x stroke, mm	73.7 x 87.4
Displacement, cc/cu in.	1493/91.0
Compression ratio	7.5:1
Bhp @ rpm, net	50 @ 5000
Torque @ rpm, lb-ft	67 @ 2500
Fuel requirement	unleaded, 91-oct
Transmission	4-sp manual
Gear ratios: 4th (1.00)	3.72:1
3rd (1.43)	5.32:1
2nd (2.11)	7.85:1
1st (3.41)	12.69:1
Final drive ratio	3.72:1

CALCULATED DATA

Lb/bhp (test weight)	40.1
Mph/1000 rpm (4th gear)	17.9
Engine revs/mi (60 mph)	3360
R&T steering index	0.70
Brake swept area, sq in./ton	232

ROAD TEST RESULTS

ACCELERATION

Time to distance, sec:	
0-100 ft.	3.9
0-500 ft	10.6
0-1320 ft (¼ mi)	20.3
Speed at end of ¼ mi, mph	69.5
Time to speed, sec:	
0-30 mph	4.4
0-50 mph	10.2
0-60 mph	14.3
0-70 mph	20.7
0-80 mph	33.4

SPEEDS IN GEARS

4th gear (4500 rpm)	85
3rd (5500)	68
2nd (5500)	47
1st (5500)	29

FUEL ECONOMY

Normal driving, mpg	28.5

BRAKES

Minimum stopping distances, ft:	
From 60 mph	189
From 80 mph	321
Control in panic stop	very good
Pedal effort for 0.5g stop, lb	38
Fade: percent increase in pedal effort to maintain 0.5g deceleration in 6 stops from 60 mph	97
Overall brake rating	fair

HANDLING

Speed on 100-ft radius, mph	33.2
Lateral acceleration, g	0.737
Speed thru 700-ft slalom, mph	50.1

INTERIOR NOISE

All noise readings in dBA:	
Constant 30 mph	72
50 mph	81
70 mph	86

SPEEDOMETER ERROR

30 mph indicated is actually	31.5
60 mph	60.5
70 mph	69.0

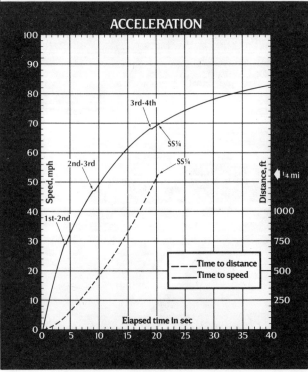

ACCELERATION

MGB

End of the line for the marque?

BY RON WAKEFIELD PHOTOS BY THE AUTHOR

IT WAS IN November 1962 that I motored southward from Detroit in my shiny green 3-year-old Austin-Healey 100-6. The Healey was my fourth British sports car, a graduation present from my father; it was still in mint condition and by far the best one so far. Unlike the two MGs and one Jaguar I'd had before it, it did not require constant mechanical coddling to keep it on the road.

But it had been outdated. I'd read in *Road & Track* about a new, *modern* sports car called the MGB. I'd been living in Detroit for a while now, and it gets beastly cold there. I was tired of bundling up to keep warm in cars with side curtains instead of roll-up glass windows. And tired of knowing in advance that, should the temperature drop to zero or so overnight, my cute little British roadster wasn't about to start on its own.

That year both the MGB, replacing the MGA, and a face-lifted Austin-Healey appeared with roll-up windows. To me, though, the Healey was already a very old-fashioned car, and even if I was pretty happy with mine I didn't want to replace it with another one. The MGB looked a lot better, even if it did have only a 4-cylinder engine and not a six like the Healey, and I liked what I read in that early R&T road test. So I called my father down in Mineola, Texas. He was a car dealer, had always gotten me good deals on my cars and surely would arrange for a new MGB at close to dealer's cost. He did, and in November I headed for Texas on my last journey with the tough old Healey.

As I recall, I paid something under $2900 for my new, bright red MGB. It was an impressive thing: contemporary of line, a lovely exhaust note from the newly enlarged British Motor Corporation "B" engine (it had reached 1.6 liters in the MGA, was now 1.8), the usual crisp, precise MG gearbox, leather seats—yes, it did have them then—and roll-up windows.

How I ever got the idea that those roll-up windows would really make all that much difference in the bitter northern cold, or that perhaps the MGB would start more readily than the three previous roadsters I'd tried valiantly to operate in Michigan, I don't know. The MGB turned out to be little warmer inside than the Healey, it didn't like starting at zero Fahrenheit any better—and to boot, its engine guzzled oil and had to be overhauled within the first year. It was my first brand-new car, and a terrible disappointment. That was my fault: I merely expected too much from it. Within a year it had been replaced by my second new car, one far better suited to Michigan cold and road salt: a Corvette.

Little did I suspect at the time that the B just might be MG's last sports car. But here it is, almost 18 years later, and MG will soon be shutting down the old MG line in Abingdon-on-Thames for good without a successor in sight. Lord knows, MG—British Motor Corporation—British Leyland—BL Limited—and whatever the conglomerate that now builds these cars has chosen to call itself with the passage of time, has lived with hard times longer than anybody cares to remember. But it is nevertheless a little hard to grasp, to believe, that the MG factory is being closed without firm plans for a new sports car carrying the name. Oh, the American dealers seem to be convincing Britain that it wouldn't be wise to let the name die, and maybe there will be a badge-engineered version of the Triumph TR7 called the MGD or some such, and maybe farther down the road there might be an open edition of the coming Leyland-Honda or some other new BL car with the fine old Morris Garages octagon. But for now, it seems that the MG name will just disappear, at least for a while.

The little MG Midget, which had evolved from the original Austin-Healey Sprite into the form tested in the last edition of the *Guide*, has been out of production for many months now. And for 1980, MGBs are being built only for the 49 U.S. states excluding California. As this goes to print, there is a supply of 1979 MGBs to last California customers until about midyear, says John Burman of JRT's San Francisco zone office, and outside California the 1980s should be available well into 1981.

Over those 18 years the B has evolved mostly in response to changing U.S. and California regulations. A GT (coupe) variation and the MGC, powered by the Austin-Healey engine after that car was discontinued in 1967, came and went. The MGB's instrument-panel design has been changed twice; various controls have been modernized, two carburetors replaced with a single one, the compression ratio dropped, two windshield wipers replaced with three—crash-safety, emission and windshield-wiping regulations were behind such changes. The lost power in 1975 as the carburetion was decimated was bad enough (about 20 bhp were lost, of which some 4½ were restored recently by minor tuning changes), but 1974's revisions to meet the new U.S. bumper regulation brought even more ignominy to the MGB.

That was the year America's National Highway Traffic Safety Administration, more or less arbitrarily, decreed that bumpers must be able to absorb the energy of a 5-mph crash into a solid, immovable barrier, and furthermore, that their heights be standardized. As far as the first requirement was concerned, the MG people did a creditable job: Instead of clumsy add-on steel or aluminum bumpers they designed nicely integrated soft-face absorbers for the B, incorporating vestiges of the traditional MG grille shape into the front one. But to meet the height requirement, they simply raised the car about 1½ in. on its suspension—probably because the two aspects of the new law were announced at different times.

The elevation job gave the formerly low-slung MGB a curious high-pockets look and turned what had been sporty, if strictly traditional, handling into a tipsy sort of behavior that must have been embarrassing to those MG people. But nothing, neither a lack of performance nor a lack of handling nor the funny new look, seemed to deter people from buying MGBs. They just kept on, and British Leyland just kept on cranking them out.

Today's MGB is almost exactly the same as the 1975 model. There's the extra handful of horsepower, but they hardly make themselves felt; there is an 85-mph speedometer to satisfy the latest whims of the NHTSA. And there is a Limited Edition trim package, which gave the test car we drove a little extra visual interest. Room has been found for a stereo radio, something that didn't even exist for cars when the B was introduced, and the various warning lights the NHTSA also requires. Remarkably, the MGB now costs nearly three times its original price—and $2000 more than two years ago.

It was a trip back in time as we motored around San Francisco and northward on the beautiful coastal highway in the B. The top had been up when we collected the car, and in this state the B is at its worst—loud, with wind and mechanical noises coming at you from all sides, and claustrophia-inducing. Put down the top—the operation is an exercise in either patience or frustration, whichever takes over in the process—and most of the mechanical cacophony disappears under the gentle rush of wind around the suddenly happy occupants. It won't be wise to try driving the B

fast or cornering it very briskly, but if one is content to just enjoy it for what it is, there is still enjoyment there. After all, who drives fast in America anyway?

As far as the engine and gearbox are concerned, the old roadster is still pretty driveable. Though the 1798-cc, 67-bhp four is controlled by a mere 2-way catalyst and has exhaust-gas recirculation, it drives surprisingly well; in fact, I've been amazed repeatedly in driving the later MGBs at how well its engine has withstood the various stages of strangulation it has been subjected to. The 4-speed gearbox, which acquired a synchronized 1st gear somewhere along the way, is mostly pleasant to operate too, though at rest it sometimes stubbornly resists being put into that gear. Which reminds me of the reason a certain British Motor Corp chief engineer used to give for the company's earlier reluctance to synchronize 1st: that we'd have trouble engaging it then. It hasn't happened with other cars, but for MG this seemed to turn into a self-fullfilling prophecy.

What else? Oh, the steering is heavy, the glovebox lock fell out on my first attempt to use it, the seatbelts are hard to find and fish out, the brakes were squishy on the test car . . .

But then it's too late to subject the MGB to the rigors of a critical road test. What I would like to do is reassure the people in power at BL Ltd, the ones trying to get that so dolorously ailing company onto its feet, that the name MG is still highly regarded, still an appropriate one to apply to a sports car. And I believe that there will continue to be a healthy market for a simple, relatively

inexpensive sports car—especially in the U.S. The MG name could be newly profitable for BL.

There's a lot of room for innovation here. Today's production car, sports or family, must meet a battery of emission, safety and fuel-economy regulations. It's not economical to produce a small sports car that differs greatly in the basic engineering from a company's own mass-produced family cars; the numbers sold, however high they may be for a sports car, don't justify the tooling costs. Certainly not if the maker is aiming at a low price.

But there's any number of small, front-drive sedans on the market today that can run circles around the MGB—items like the Honda Civic, Mitsubishi (Dodge) Colt, Volkswagen Rabbit and Ford Fiesta come to mind. The same company that built the MG even has a new one coming, the Mini Metro, and logically that would be the right basis for a new MG. Or the Honda-Leyland. The engineering hardly needs to vary from the sedan; the roadster body, however, does. But it could be simple, offering an absolute minimum of amenities and some cost-saving innovations, such as body panels that are interchangeable side-to-side and front-to-rear. Pininfarina showed how to do this a few years ago with the Peugette, a terrific little 2-seater incorporating these ideas on the front-drive Peugeot 104 chassis.

Somebody is going to offer such a car, I believe—if it's not an MG, it very well may be a Ford, or a Honda, or even a Peugeot. And when that somebody does, the college students of the Western world will beat a path to his door.

PRICE

List price, all POE	$7950
Price as tested	$8900

Price as tested includes Limited Edition package ($600), AM/FM stereo radio ($200), dealer prep (est $150)

GENERAL

Curb weight, lb/kg	2335	1059
Test weight	2505	1136
Weight dist (with driver), f/r, %	52/48	
Wheelbase, in./mm	91.1	2314
Track, front/rear	49.0/49.2	1245/1250
Length	158.2	4018
Width	59.9	1521
Height	51.0	1295
Trunk space, cu ft/liters	2.9	82
Fuel capacity, U.S. gal./liters	13.0	49

ENGINE

Type	ohv inline 4
Bore x stroke, in./mm	3.16 x 3.50 ....80.3 x 89.0
Displacement, cu in./cc	110 1798
Compression ratio	8.0:1
Bhp @ rpm, SAE net/kW	67/50 @ 4900
Torque @ rpm, lb-ft/Nm	94/127 @ 2500
Carburetion	one Zenith 150CD4T
Fuel requirement	unleaded, 91-oct

DRIVETRAIN

Transmission	4-sp manual
Gear ratios: 4th (1.00)	3.89:1
3rd (1.38)	5.37:1
2nd (2.17)	8.44:1
1st (3.44)	13.38:1
Final drive ratio	3.89:1

CHASSIS & BODY

Layout	front engine/rear drive
Body/frame	unit steel
Brake system	10.75-in (273-mm) discs front, 10.0-in. (254-mm) drums rear; vacuum assisted
Wheels	cast alloy, 14 x 4½J
Tires	Uniroyal Rallye 180/70, 185/70SR-14
Steering type	rack & pinion
Turns, lock-to-lock	2.9
Suspension, front/rear: unequal-length A-arms, coil springs, lever shocks, anti-roll bar/live axle on leaf springs, lever shocks	

CALCULATED DATA

Lb/bhp (test weight)	37.4
Mph/1000 rpm (4th gear)	18.6
Engine revs/mi (60 mph)	3225
R&T steering index	0.93
Brake swept area, sq in./ton	248

ROAD TEST RESULTS

ACCELERATION

Time to distance, sec:

0-100 ft	3.6
0-500 ft	10.3
0-1320 ft (¼ mi)	19.6
Speed at end of ¼ mi, mph	69.5

Time to speed, sec:

0-30 mph	4.0
0-50 mph	9.3
0-60 mph	13.6
0-70 mph	20.3
0-80 mph	31.5

SPEEDS IN GEARS

4th gear (5000 rpm)	94
3rd (5500)	74
2nd (5500)	49
1st (5500)	32

FUEL ECONOMY

Normal driving, mpg	18.5

BRAKES

Minimum stopping distances, ft:

From 60 mph	175
From 80 mph	320
Control in panic stop	fair
Pedal effort for 0.5g stop, lb	25
Fade: percent increase in pedal effort to maintain 0.5g deceleration in 6 stops from 60 mph	60
Overall brake rating	fair

HANDLING

Lateral accel, 100-ft radius, g	est 0.71
Speed thru 700-ft slalom, mph	est 54.0

INTERIOR NOISE

Constant 30 mph, dBA	70
50 mph	75
70 mph	85

SPEEDOMETER ERROR

30 mph indicated is actually	31.0
60 mph	60.0

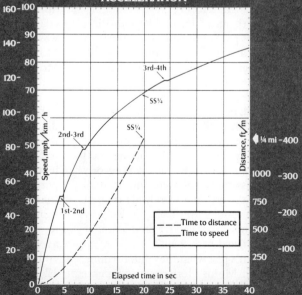

ACCELERATION

BRITISH MOTOR HERITAGE

MGB

New hope for the MG America loved last

BY PETER EGAN

HERE YOU ARE, owner of a 1968–1973 MGB, mourning the fate of your car. Oil changes and tuneups have kept the engine running fine, the all-synchro gearbox is still good, the brakes stop, the shocks damp and the steering steers.

Unfortunately, all these components are bolted onto a rusted-out body. Heroic efforts—welding and new steel panels—have probably kept the car on the road for a while, but one day you are forced to admit that nothing solid remains to which panels can be welded. You look at your faithful MGB and realize that what you have in your garage is, essentially, a parts car. So you strip off the valuable pieces and send the chassis to the crusher. Your old MG, with its history, title and registration numbers, is gone forever.

With more than a half-million MGBs produced from 1962 through 1980 and 130,000 of those cars still registered in the U.S., this has become a common scenario and a sad end to more than a few of these cars. So sad, in fact, that two Englishmen, Peter Mitchell and David Bishop, finally decided to do something about it. They discovered that about 1000 tons of original MG factory press dies and assembly jigs were still around. With the growing worldwide interest in the restoration of old MGBs, why not (they asked) revive production of the original bodyshells?

That's just what they did, forming British Motor Heritage in coopera-

tion with Rover Group. The old tools were uncrated and refurbished, an assembly line was laid out at Faringdon in Oxfordshire, workers were hired (many from the old MG days), and new bodyshells began rolling off the line, those for export complete with doors, hood, fenders and trunklid. The result was not a replica body, but an exact replacement for the original, built on the factory equipment, to factory standards, right down to the grade of steel and the type and number of welds.

To publicize the new American version, BMH imported a tired old 1972 MGB from the U.S., used it as a parts (and title) donor for a new BMH bodyshell and had a crew of four work around the clock to reconstruct a new car in just 16 days. They then air-freighted the MG to New York and drove it 3000 miles from coast to coast. This new/old MGB was left as a demonstrator with Moss Motors, Ltd. in Goleta, California. Moss is a specialist supplier of MG and other British parts, and a designated BMH dealer.

Ken Smith, public relations coordinator at Moss, invited us to borrow the red car for a road test, so I flew up to Santa Barbara, just south of Goleta. Smith picked me up at the airport in his own MGB and gave me a tour of the Moss facility, whose vast warehouse of British parts is enough to arouse latent kleptomania in the average English car buff. At the far end of the warehouse were stacked 12 new BMH bodyshells, in gray

primer. We wheeled one out into the sunlight for a better look.

Still a lovely shape after all these years, the complete shell sells for $3995 in the U.S. It is intended as a replacement for the chrome-bumpered (pre-rubber-nosed) cars built from 1968 through 1973, though owners of pre-1968 cars have reported that the earlier transmission and starter can be installed with minor modification (a few well-placed hammer blows) to the transmission tunnel. All chassis are coated with electropheretically applied primer, and all holes are drilled, except those for the trim strip on the door, which has to be aligned after final door-hinge adjustment.

The bodyshell looks so clean and immaculate that it's hard to imagine anyone transferring worn or dirty parts from an old donor car. The natural instinct will be to refurbish the old parts and buy a lot of new ones—an instinct on which both Moss and Rover Group are counting.

The British Motor Heritage car we tested was assembled by the MG specialist firm of Brown & Gammons in England. The car has chrome bumpers, two SU carburetors, an all-synchro 4-speed transmission with overdrive, chrome wire wheels, a Moto-Lita steering wheel, flawless red paint and a Moss black leather interior with red piping. The engine is a stock 1.8-liter unit, except for a mild street grind on the cam, and the detailing of hoses, lines and ancillary items in the engine compartment is of show quality.

PHOTOS BY RICHARD M. BARON

On the drive down the coast to Newport Beach, the car's mechanical integrity proved to be as good as its appearance. While I've never driven a brand-new MGB (being on my third used one), the BMH car feels as I imagine a showroom-new 1972 MGB must have felt. New Armstrong lever dampers, taut steering, fresh springs and new suspension bushings all work to accentuate the traditional, basic charm of the MGB, which is a feeling of solidness and mechanical honesty.

Ride is excellent over a variety of road surfaces—not quite as harsh or stiff as our shorter-wheelbase Miata, which drew many natural comparisons from the staff during our test. Steering is slightly heavy by current standards, but with good road feel through the wheel. In hard cornering the car initially exhibits a fair amount of body roll, then takes a set and motors through predictably, with mild understeer. Grip is good with the Uniroyal Rallye 185/70HR-14 tires, the car generating a creditable 0.80g of lateral load on the skidpad.

One of the most pleasant aspects of the BMH car is its overdrive. Flick the wiper stalk toward you in 3rd or 4th gear and the Laycock de Normanville overdrive unit gently shifts up, dropping highway revs to a less hectic level. In 4th gear at 65 mph, the tach falls from 3400 rpm to about 2700, and the relaxation factor in losing those 700 revs is astounding. The BMH MGB drones along without commotion at 75 mph, while my own 1973 roadster, without overdrive, sounds as if it's consuming its innards at that speed.

Instruments are nicely laid out: round faces, all visible through the steering-wheel rim, with a rectangular oil-pressure gauge in the center. Passenger leg room is so good that short people can't reach the end of the footwell, while the driver's side is about right for this 6-ft. 1-in. driver with the seat all the way back, at full rear tilt. With the top up, the windshield seems rather short, and you have the sense of driving along with a cap pulled down low over your eyes. Visors are almost redundant. Rotating heater controls are the usual stiff, dumb design, and they often work when they are new. These did.

Cold starting is done with the lock-

■ Better than 1972? First-rate assembly and attention to detail on the BMH car by Brown & Gammons Ltd. have set a high standard for the home mechanic.

BRITISH MOTOR HERITAGE MGB

0–60 mph	12.8 sec
0–¼ mi	19.0 sec
Top speed	est 105 mph
Skidpad	0.80g
Slalom	na
Brake rating	average

PRICE

List price for bodyshell .... **$3995**   Price as tested ... **est $12,000–$20,000**

Price as tested will vary greatly depending on the level of restoration you choose and how much of the work you do yourself.

ENGINE

Type	ohv **inline-4**
Displacement	111 cu in./1798 cc
Bore x stroke	3.16 x 3.50 in./ 80.3 x 89.0 mm
Compression ratio	8.8:1
Horsepower (SAE):	**92 bhp @ 5400 rpm**
Torque	**110 lb-ft @ 3000 rpm**
Maximum engine speed	6000 rpm
Fuel injection	two (1V) SU carburetors
Fuel	prem leaded, 91 pump oct

DRIVETRAIN

Transmission			**4-sp manual with OD**
Gear	Ratio	Overall ratio	(Rpm) Mph
1st	3.44:1	13.45:1	31
2nd	2.17:1	8.48:1	50
3rd	1.38:1	5.40:1	78
3rd OD	1.38:1	4.31:1	97
4th	1.00:1	3.91:1	est (5835) 105
4th OD	1.00:1	3.13:1	est (4670) 105
Overdrive ratio			0.80:1
Final drive ratio			3.91:1
Engine rpm @ 60 mph in 4th OD			2670

CHASSIS & BODY

Layout	**front engine/rear drive**
Body/frame	unit steel
Brakes, f/r	**10.8-in. discs/10.0 x 1.7-in. drums,** vacuum assist
Wheels	wire spoke, **14 x 4½J**
Tires	Uniroyal Rallye 340/70, **185/70HR-14**
Steering	**rack & pinion**
Turns, lock to lock	2.9
Suspension, f/r:	**upper & lower A-arms,** coil springs, lever shocks, anti-roll bar/ **live axle,** leaf springs, lever shocks

FUEL ECONOMY

Normal driving	est 25.0 mpg
EPA city/highway	na
Fuel capacity	12.0 gal.

INTERIOR NOISE

Idle in neutral	63 dBA
Constant 70 mph	83 dBA

GENERAL DATA

Curb weight	**2305 lb**
Test weight	**2450 lb**
Weight dist, f/r, %	**51/49**
Wheelbase	91.0 in.
Track, f/r	49.3 in./49.3 in.
Length	152.7 in.
Width	59.9 in.
Height	49.8 in.
Trunk space	2.9 cu ft

ACCELERATION

Time to speed	Seconds
0–30 mph	4.3
0–60 mph	12.8
0–80 mph	24.0
Time to distance	
0–100 ft	3.8
0–500 ft	10.3
0–1320 ft (¼ mi)	19.0 @ 72.5 mph

BRAKING

Minimum stopping distance	
From 60 mph	170 ft
From 80 mph	300 ft
Control	very good
Pedal effort for 0.5g stop	35 lb
Fade, effort after six 0.5g stops from	
60 mph	45 lb
Brake feel	very good
Overall brake rating	average

HANDLING

Lateral accel (200-ft skidpad)	0.80g
Balance	moderate understeer
Speed thru 700-ft slalom	na
Balance	na

Subjective ratings consist of excellent, very good, good, average, poor; na means information is not available.

---

ing choke cable fully pulled, and the knob has to be eased in by stages until the engine is warm. While acceleration is leisurely by modern standards (0–60 mph in 12.8 seconds), the engine accelerates with a pleasant deep, mellow note. Midrange torque is excellent, giving the car real-world driveablity in traffic maneuvers. Brakes are medium good, if not world-class, stoppers, requiring more leg muscle than most current cars.

Everyone on the staff who drove the MGB was impressed and charmed by it, and more than one found himself wondering if a new/used MGB might not be a rational alternative to a brand-new current sports car. Combine a rusty, used MGB with a new $3995 British Motor Heritage bodyshell, and you have a project starting point of around $5500.

With paint, wheels, top, tires, chrome, interior kit, a drivetrain rebuild and a few careless moments with the Shiny New Parts Catalog, it's not hard to imagine that an owner could spend another $10,000 completing the project, not to mention some long hours of labor.

Which makes the brand-new rebuilt MGB—as several people have pointed out—about as expensive as a new Miata. Neither car, however, is really a replacement for the other. A Miata is a modern, competent sports car that you can buy now and drive hard with little need for hands-on mechanical involvement. A British Motor Heritage MGB is a labor of love that just happens to work quite nicely as an automobile. It is also a sports car whose character and subtle charms have to be experienced before you can understand why so many people have worked with such enthusiasm to save it.

Test Notes . . .

■ Dropping the clutch for our timed acceleration runs brings a brief chirp from the tires and usable power up to about 5000 rpm, though the engine's mechanical noise is high by today's standards.

■ Around the skidpad, the MGB generated 0.80g—a surprisingly high level of lateral acceleration—together with moderate understeer that's greatly amenable to throttle and steering.